CUDA 13

High-Performance Computation and Graphics

Matthew Scarpino

CUDA 13

High-Performance Computation and Graphics

Matthew Scarpino

Printed in the United States of America

Publisher's Cataloging-in-Publication data

Author: Scarpino, Matthew

Title: CUDA 13: High-Performance Computing and Graphics

Description: Explains how to write high-performance applications using NVIDIA's CUDA framework

ISBN: 978-0-9973037-4-2

Table of Contents

Chapter 1

Introduction to CUDA Development

I wrote my first GPU kernel in 2010, about six months after the initial release of OpenCL. I was deeply impressed with the processing speed, but I never thought general-purpose GPU computing would become popular. Users wanted faster rendering from their graphics cards, not *math*.

But then Bitcoin came along, and people quit their jobs to mine cryptocurrencies. Then high-speed machine learning became insanely important. As I write this in 2026, customers are paying several thousand dollars to buy cards powerful enough to run large language models (LLMs). The global GPU market is projected to reach $325 billion by 2031. Frame rate? Texture memory? Who cares?

Just as GPUs have become the dominant devices for computing, CUDA has become the dominant language for harnessing them. Every major toolset for machine learning, from PyTorch to TensorFlow to JAX, can use CUDA for accelerated computation. CUDA is also popular in libraries for scientific computing, video processing, and computer vision. CPUs can still get the job done, but CUDA will get the job done *fast*.

In August 2025, Nvidia released CUDA 13.0, the first major release in three years. As you might expect, the focus is on accelerating the matrix operations used in machine learning. This book discusses these advanced matrix operations as well as traditional CUDA operations, including frequency analysis, image processing, solving systems of equations, and graphical acceleration.

In my opinion, the best way to learn CUDA is to work with practical code. Understand the functions and data structures first, then walk through the compilation and execution of examples.

With this in mind, each chapter has one or more projects that demonstrate how theory can be put into practice. Later in this chapter, I'll explain where to download the example code for your operating system.

Before we get into technical details, I'd like to provide a brief overview of CUDA and its evolution. Nvidia has spent an extraordinary amount of time developing this language, and once you see why, you'll have a better idea why learning about CUDA is worth your own time.

NOTE:

Machine learning is a central priority in modern computing, but it's not one of the topics explored in this book. This book focuses on the CUDA Toolkit, which is introduced in this chapter and discussed in detail in Chapter 2. For information about using CUDA for machine learning, visit https://developer.nvidia.com/cudnn.

1.1 A Brief History of CUDA

Every year, the Association for Computing Machinery (ACM) hosts a conference for graphics enthusiasts called SIGGRAPH (Special Interest Group on Computer GRAPHics). At SIGGRAPH 2004, Ian Buck and other Stanford researchers presented a programming language intended to execute stream computing tasks on graphics cards. Their stated goal was to provide "GPU-based computing for the masses".

Their language was called Brook, and they demonstrated how simple functions called *kernels* could run on a GPU and perform large-scale operations in parallel. An example kernel is given as follows:

```
kernel void foo (float a<>, float b<>, out float result<>) {
  result = a + b;
}
```

The presentation explained how kernels like this could be used not only for basic math, but also for ray tracing, frequency analysis, and image processing. When they compared the performance of Brook against that of a Pentium 4, the results made it clear that GPU computing was a field worth pursuing.

When Ian Buck left Stanford, he joined Nvidia and worked with John Nickolls, the director of GPU computing. Together, they took the kernel-based execution methodology and created CUDA 1.0, released in June of 2007. Table 1.1 lists CUDA 1.0 and successive versions of the language, up to the latest version.

Table 1.1
CUDA History

Version	Year	Description
1.0	2007	Executes C-like functions called kernels
2.0	2008	Support for atomic operations, double-precision floating-point values, and three-dimensional textures
3.0	2010	Support for the Fermi architecture, support for fused multiply-add (FMA) operations, interoperability with Direct3D and OpenGL
4.0	2011	Ability to share GPUs across multiple threads, use of pinned memory on the host, support for inline PTX (Parallel Thread Execution)
5.0	2012	Dynamic parallelism, support for __device__ functions, integration of a command-line profiler
6.0	2014	Support for the Maxwell architecture and unified memory
7.0	2015	Support for separate streams of execution, improved context management
8.0	2016/2017	Support for GPUDirect Async and NVLink high-speed interconnect
9.0	2017	Support for the Volta architecture and cooperative groups, improved support for half-precision floating-point values
10.0	2018	Support for the Turing architecture, CUDA graphs, warp functions, and interoperability with Vulkan
11.0	2020	Support for the Ampere architecture, asynchronous data transfer, and improved memory management
12.0	2022	Support for the Hopper and Ada Lovelace architectures, cooperative grid arrays (CGAs), and asynchronous transaction barriers
13.0	2025	Support for Blackwell and Jetson devices, tile-based programming, improved access with Python

Of course, the evolution of CUDA has been driven in large part by the evolution of Nvidia's hardware. As graphics cards acquired new capabilities, CUDA expanded to allow developers to take advantage of them.

For example, Nvidia released the Fermi architecture in 2010 in its effort to pursue the high-performance computing (HPC) market. Fermi devices provide many features demanded by HPC, including fused multiply-add (FMA) instructions and 32 cores dedicated to CUDA. After making Fermi devices available, Nvidia released CUDA 3.0, which supported newer math instructions (including FMA), 64-bit data types, and concurrent kernel execution.

To respond to the growing interest in deep learning, Nvidia developed the Volta architecture. These devices were the first to contain *tensor cores*, which accelerate the matrix-multiply-accumulate (MMA) operations that take up so much time in machine learning. To enable developers to access tensor cores, CUDA 9.0 was expanded to provide the Warp Matrix Multiply Accumulate (WMMA) API, which will be discussed at length in Chapter 5.

Despite the many changes in CUDA's capabilities and libraries over the years, the fundamentals of CUDA development haven't significantly changed since Ian Buck gave his presentation at SIGGRAPH 2004. The following section presents a high-level overview of the coding process.

1.2 Overview of CUDA Development

CUDA is a versatile framework that supports different languages, but this book focuses on writing CUDA applications in C++. For C++ developers, there are three main ways to code applications with CUDA:

1. **The Easy Way** — call functions in CUDA libraries
2. **The Hard Way** — write custom CUDA kernels
3. **The Really Hard Way** — write code in the virtual assembly language, PTX (Parallel Thread Execution)

This section looks at each of these methods and presents an overview of how applications are coded. Later chapters will expand on this in detail.

1.2.1 CUDA Libraries

Most of the chapters in this book explain how to access the libraries provided by Nvidia. If you have a choice between calling a function in a CUDA library and writing a custom kernel, I strongly recommend calling the library function.

Table 1.2 lists eleven of the libraries available. The second column presents the chapter in which the library is discussed. If the second column contains --, it means that library won't be discussed in this book.

For example, Chapter 6 discusses the cuBLAS library, whose functions perform operations related to linear algebra. Chapter 7 presents the cuSOLVER library and the cuDSS library, whose functions make it possible to solve systems of equations.

Table 1.2
Important CUDA Libraries

Library	Chapter	Description
cudart	3, 4, 5	CUDA runtime library - provides the basic C/C++ interface for CUDA development
cuBLAS	6	Provides functions for linear algebra
cuSOLVER	7	Provides functions that solve systems of equations represented by dense matrices
cuDSS	7	Provides functions that solve systems of equations represented by sparse matrices
cuFFT	8	Provides functions that analyze frequencies using the Fast Fourier Transform (FFT)
NPPI	9	Provides functions that analyze and transform images (NPPI stands for NVIDIA Performance Primitives Image)
cuRAND	--	Provides functions that generate random numbers
nvJPEG	--	Provides functions that encode and decode JPEG images
NVML	--	Provides functions that monitor and manage hardware at the system level
libcuda	App A	Provides functions that access the CUDA Driver API, providing low level access to Nvidia's hardware
NVRTC	App A	Provides functions that load and compile CUDA code programmatically

Most of these libraries are provided inside Nvidia's free CUDA Toolkit. Chapter 2 is devoted to this toolkit, and it starts by explaining how to download and install it. Then it explores the utilities and libraries provided.

The largest and most important library in the toolkit is the CUDA runtime library. This is required in the vast majority of CUDA applications, and is needed to make CUDA accessible in C/C++ applications. Chapters 3, 4, and 5 present the many functions provided by this library.

Working with matrices is a major concern in graphics and machine learning, and CUDA provides multiple libraries that perform matrix operations. One popular set of routines is called BLAS (Basic Linear Algebra Subroutines), and Nvidia provides its implementations of these routines in the cuBLAS library.

For systems of equations involving matrices, Nvidia provides the cuSOLVER library for dense matrices and the cuDSS (Direct Sparse Solver) library for sparse matrices. cuDSS is so new that it needs to be downloaded separately from the CUDA Toolkit.

The last two libraries in the table are particularly interesting. The CUDA Driver API provides many low-level capabilities for accessing Nvidia's hardware. One capability involves deploying code to the hardware. By combining its functions with those of the Nvidia Runtime Compilation (NVRTC) library, applications can dynamically load, compile, and deploy code to a graphics card. Appendices A and B discuss this advanced topic and show how it can be implemented in code.

The process of calling library functions in CUDA consists of four steps:

1. Allocate memory on the device to hold input data and processed data.
2. Copy input data from the host to the device.
3. Process the input data by running one or more functions from the CUDA library.
4. Copy the processed data from the device to the host.

For example, suppose you want to use cuBLAS to multiply Matrix A by Vector x to obtain Vector y. You'd call `cudaMalloc` three times to allocate memory on the device for A, x, and y. Then you'd call `cudaMemcpy` twice to copy input data to A and x.

Once the data has been transferred, the `cublasSgemv` function will perform matrix-vector multiplication on the target device. Once that's finished, you'd call `cudaMemcpy` again to copy the result to the host. Chapter 6 provides a full exploration for cuBLAS, but the overall process really is quite simple.

1.2.2 Writing Custom Kernels

If you can't find a suitable function in a CUDA library, you can write a custom kernel that will execute on the device. Kernels look like regular C functions, with function names followed by parameters in parentheses and instructions in curly braces. For example, the kernel in the ch04_add_arrays project is given as follows:

```
__global__ void addArr(const float* a, const float* b, float* c) {

    // Determine the Thread ID
    int i = threadIdx.x;

    // Add elements of a and b
    c[i] = a[i] + b[i];
}
```

Even if you don't understand every word, it should be clear that the kernel adds elements of the `a` and `b` arrays and stores the sums in the `c` array. Each array was allocated using `cudaMalloc` and the `a` and `b` arrays were copied to the device with `cudaMemcpy`.

To write kernels that take full advantage of the hardware, you need to have a solid understanding of three topics:

1. **Memory spaces** — when to store data in global memory, shared memory, constant memory or registers
2. **Execution model** — splitting work into threads, organizing threads into thread blocks, arranging thread blocks into grids
3. **Synchronization** — ensuring processing continues only after the required data has been processed and stored

These topics aren't simple, but once you understand how kernels execute, you'll be able to harness the parallelism provided by GPUs. Chapter 4 explores the topic of kernel development and presents several examples of how they can be coded.

1.2.3 Parallel Thread Execution (PTX)

For those really interested in low-level programming, Nvidia provides a virtual assembly language called PTX (Parallel Thread Execution). PTX is similar to a regular assembly language, but you don't need to know details of the target device, such as the number of registers. Instead, you can write PTX code and run it on many different types of devices. This is why PTX is referred to as *virtual.*

Like regular assembly language, PTX consists of opcodes followed by parameters. For example, the following code implements a portion of a traditional `for` loop:

```
// Check if loop should end
setp.ge.u32  %loop_done, %i, %n;
@%loop_done bra loop_end;

// Read next value
mul.wide.u32  %in_offset, %i, 4;
add.u64       %current_addr, %in_addr, %in_offset;
ld.global.u32 %current_val, [%current_addr];
```

Most CUDA developers stay away from PTX, but there are two good reasons to learn the language. First, PTX receives new instructions before the CUDA runtime library, so PTX lets you work with the latest and greatest operations.

Second, the Nvidia Runtime Compilation (NVRTC) library makes it possible to programmatically compile PTX to device code. By combining this with the Driver API, applications can dynamically generate PTX, compile it, and deploy it to a device. Appendix A discusses NVRTC and the Driver API, while Appendix B presents PTX in its complex glory.

1.3 Book Structure

CUDA is a vast subject, and this book doesn't touch on every library and utility. Instead, the content focuses on four topics:

1. **Fundamentals** — tools in the CUDA Toolkit, functions of the CUDA Runtime, coding custom kernels, advanced features
2. **Accessing libraries** — calling functions from CUDA libraries like cuBLAS, cuSOLVER, cuDSS, cuFFT, and NPP/NPPI
3. **Graphics acceleration** — rendering three-dimensional graphics with OpenGL and accelerating graphics with CUDA
4. **JIT (Just-In-Time) compilation** — coding applications in PTX and compiling and deploying the code programmatically

This section provides an overview of these topics and shows how they relate to the book's chapters and appendices.

1.3.1 CUDA Fundamentals

If you're just starting out, the first step is to download the CUDA Toolkit from Nvidia and become familiar with its files and utilities. Chapter 2 covers this in detail and walks through the process of using the Nvidia compiler (`nvcc`) to build a simple CUDA application.

Once you understand the tools, you'll be ready to look at code. Chapter 3 introduces the basic functions of the CUDA Runtime, which perform operations like allocating memory on the device and transferring data between the host and device.

Chapter 4 is concerned with kernels: coding, compiling, and executing. This is one of the most complicated topics in this book, so if you don't need to write kernels, you may want to skip it. Writing a kernel function isn't particularly difficult, but it can be hard to keep track of the different types of memory (global, constant, shared, register) as well as the terminology for managing threads (thread block, grid, warp).

Chapter 5 presents features of CUDA that are interesting but not absolutely necessary. For example, using streams and pinned memory on the host can improve performance, but applications will run fine without them. Dynamic parallelism is a fascinating feature that makes it possible to launch one kernel from another kernel, but most developers don't use it. The most interesting topic discussed in Chapter 5 is the Warp Matrix Multiply Accumulate (WMMA) API, which accelerates the process of multiplying two matrices and adding a third.

1.3.2 Accessing Libraries

If you can find a library function that will perform your task, you're much better off calling the library function than writing a custom kernel. Chapters 6 through 9 discuss many of the libraries provided in the CUDA Toolkit, including cuBLAS, cuSOLVER, and the Nvidia Performance Primitives (NPP).

Chapter 6 shows how cuBLAS can be used to perform basic linear algebra operations, such as dot products, matrix-vector multiplication, and matrix-matrix multiplication. Chapter 7 expands on this and shows how to solve systems of equations. To be specific, it shows how to use cuSOLVER to solve equations represented by dense matrices and cuDSS to solve equations represented by sparse matrices.

Chapter 8 explains the theory of the Fast Fourier Transform (FFT), which plays an important role in the field of signal processing. Then it explains how to use Nvidia's cuFFT to compute forward and inverse FFTs using CUDA. Chapter 9 shows how Nvidia Performance Primitives for Images (NPPI) makes it possible to filter images.

1.3.3 Graphics Acceleration

Believe it or not, there are still developers using graphics cards for graphics. CUDA doesn't provide its own library for graphical rendering, so this book focuses on OpenGL, which is discussed in Chapter 10. This discussion explains the graphics pipeline and shows how shaders make it possible to customize the rendering process.

To accelerate rendering, CUDA provides a fascinating capability called OpenGL-CUDA interoperability. This makes it possible to process graphics data on the device using CUDA before it's rendered using OpenGL. Chapter 11 introduces interoperability and Chapter 12 shows how it can be used to process textures.

1.3.4 Just-In-Time (JIT) Compilation

One of the most interesting features of CUDA is that applications can generate, compile, and deploy PTX code at runtime. To see how this works, you need a solid understanding of the CUDA Driver API, Nvidia Runtime Compilation (NVRTC), and PTX coding. Appendix A discusses the first two topics and Appendix B presents the third.

While assembly language scares many developers, PTX is good to learn for many reasons. First, the better you understand low level operations, the better you can write high-level code. Second, PTX is updated more frequently than the C++ API, and Appendix B explains how to access the new WGMMA (Warp Group Matrix Multiply and Accumulate) API, which makes it possible to perform matrix operations at high speed.

1.4 Example Code

The best way to learn CUDA is to get your hands dirty and work with example code. For this reason, every chapter and appendix of this book has at least one example project. The code is freely available on Github, and the repository's address depends on your operating system:

- For Windows users, the example code is provided as a Visual Studio solution at https://github.com/mattscar/cuda-book-win.
- For Linux users, the example code is provided as a CMake project at https://github.com/mattscar/cuda-book-gnu.

Once you've downloaded the repository as a ZIP file, you can look through the projects available. Most of these projects require that you have the CUDA Toolkit installed and that its path is given by the `CUDA_HOME` variable. Chapter 2 explains how to download and install this toolkit.

A few projects have dependencies that go beyond the CUDA Toolkit. One example project in Chapter 7 requires the cuDSS package, which makes it possible to solve systems of equations involving sparse matrices. The projects in Chapters 10, 11, and 12 require the installation of OpenGL, GLUT, and GLEW.

1.5 Summary

When I first started coding general-purpose GPU applications, I used a toolset called OpenCL. This provides a basic set of capabilities and runs on multiple devices and operating systems. As much as I hate to admit it, CUDA surpasses OpenCL in every respect. Better libraries, better utilities, and far better performance.

This chapter has provided a whirlwind tour of CUDA, starting with its history and leading to its current adoption. Simply put, CUDA makes it possible for applications to harness the extraordinary power of Nvidia's GPUs. The chapters of this book will show you how to use this power to perform numeric operations and accelerate graphics.

This chapter has also presented the three basic ways to code CUDA applications. If you can find a library that meets your needs, it's a good idea to call the functions of that library. If not, you'll have to code a custom kernel and manage its threads and memory access. If you enjoy low-level development, you can write code in PTX and compile it on the fly. But remember: with great power comes great complexity...

Chapter 2

The CUDA Toolkit

Most of this book focuses on the capabilities provided by Nvidia's CUDA Toolkit, which can be freely downloaded online. Therefore, before you can start coding and building projects, you'll need to download and install the toolkit on your system. The first section of this chapter walks through the installation process for the Windows and Linux operating systems.

Once you've installed the toolkit, it's a good idea to look through the top-level directory to explore its content. For this book, the most important folders are bin, include, and lib. The bin folder includes the toolkit's utilities, the include folder contains header files that declare the toolkit's functions, and the lib folder contains the libraries that provide the functions' code.

The most important utility in the toolkit is the Nvidia compiler, `nvcc`, which manages the process of transforming source code into executables that can access Nvidia's GPUs. A large part of this chapter explains how to access `nvcc` from the command line and constrain the build process using flags.

After discussing the CUDA build process, this chapter presents a basic example of how `nvcc` can be used to build a trivial example project. Once you have this example working, you'll be ready to tackle the more interesting projects in later chapters.

The chapter ends by explaining how to download, install, and build the example code for this book. The instructions depend on whether you use Visual Studio for development or GNU tools like gcc and g++. This book takes a hands-on approach to presenting CUDA, and a large part of the content is devoted to walking through this code. For this reason, it's critical to have the example projects working correctly.

2.1 Downloading and Installation

This section explains how to download and install the CUDA Toolkit. But first, it's important to make sure that you have an Nvidia GPU connected to your system and a suitable driver for accessing the device.

2.1.1 Updating the Driver

A driver is a piece of software that enables applications to communicate with hardware. Before you can install the CUDA 13 Toolkit, you'll need to have at least version 580 of Nvidia's driver installed on your system. In general, it's always a good idea to make sure you have the latest driver installed.

There are a few ways to update an Nvidia driver. First, you can access the download page for Nvidia's drivers by visiting Nvidia's main site, https://www.nvidia.com, and clicking the Drivers link in the upper right. Then select the appropriate product category, product series, and product corresponding to your GPU. Then click the Find button and download the driver file to your system.

Another method is to download the Nvidia App to your system, and let the application select and install your driver. In the United States, the Nvidia App page is https://www.nvidia.com/en-us/software/nvidia-app. For other nations, replace en-us with the appropriate locale code, such as fr-fr, de-de, ja-jp, or zh-cn.

The Nvidia App will check the version of the driver installed on your system. If the latest version isn't installed, it will suggest driver versions. I recommend selecting the latest version available.

2.1.2 Installing the CUDA Toolkit

Once you've installed a current Nvidia driver, you're ready to install the CUDA Toolkit. The installation process depends on your operating system:

1. Open a browser and visit Nvidia's CUDA download page at https://developer.nvidia.com/cuda-downloads.
2. If you haven't already, you may need to subscribe to Nvidia's CUDA Toolkit Updates. This requires your first name, last name, email address, and location. You can unsubscribe later.
3. On the download page, select your operating system, processor architecture, operating system version, and installer type. The local installer is larger but results in faster installation.

4. On Linux, follow the instructions on the web site to install the CUDA Toolkit.
5. On Windows, click the Download button. When the download is finished, launch the executable to start installation. Accept the license agreement, choose Express installation, select the installation location, verify the environment variable setting, and then wait for installation to complete. Then click the Next and Close buttons to complete the installation process.

When installation is finished, you should find the CUDA Toolkit files in the selected location. On my Windows system, they're in C:\Program Files\NVIDIA GPU Computing Toolkit\CUDA\v13.1. On my Linux system, they're in /usr/local/cuda.

2.1.3 Verifying Installation

After you've updated the driver and installed the toolkit, it's a good idea to verify that everything's ready. You can check that the Nvidia compiler is available by running the following command on the command line:

```
nvcc --version
```

This should print a message that identifies the compiler's version and build date. Make sure that the version is 13 or higher. If it's not, you may need to update your environment variables to point to the new toolkit.

Another way to check installation is to use Nvidia's System Management Interface (SMI), which is provided in the Nvidia driver. This utility displays information about the installed driver, the installed CUDA version, and the state of your GPU. You can launch it by running the following command on the command line:

```
nvidia-smi
```

For now, it's just important to check that the driver version and CUDA version are correct.

2.1.4 Sample Code

Nvidia used to provide sample code in the CUDA Toolkit, but now the samples have to be downloaded separately. You can download the official sample code by opening a browser and visiting https://github.com/NVIDIA/cuda-samples. Then click the green button labeled Code and select Download ZIP.

2.2 Exploring the Toolkit

After the installation is finished, the toolkit's top-level folder should be assigned to the `CUDA_HOME` environment variable on your system. If you look through this folder, you'll see that it contains several folders and files, including release notes and the distribution license. For this book, we'll be concerned with the three folders listed in Table 2.1.

Table 2.1

Important Folders in the CUDA Toolkit

Folder Name	Content
bin	Tools that build and analyze executables
include	Header files needed for the build process
lib	Libraries that provide CUDA routines

This section provides a brief overview of these folders. To be specific, I'll present the utilities in the bin folder, the library files in the lib folder, and the header files in the include folder.

2.2.1 Utilities

The installation should add the toolkit's bin folder to your system's `PATH` variable. This makes it possible to access the Nvidia compiler and other utilities from the command line. Table 2.2 lists eleven of the utilities in the bin folder.

Table 2.2

CUDA Toolkit Utilities

Utility Name	Description
bin2c	Converts binary files into C-style arrays
ctadvisor	Provides advice for reducing compilation time
cu++filt	Decodes mangled C++ identifiers into readable names
cudafe++	Separates CUDA-specific code from regular C/C++ code
cuobjdump	Displays content of binary files
fatbinary	Combine CUDA objects into a single binary
nvcc	Manages the Nvidia compile process

nvdisasm	Converts (disassembles) CUDA binary files into readable text
nvlink	Links compiled CUDA device code into a binary
nvprune	Prunes object files and libraries
ptxas	Assembles source code into the PTX format

Of these utilities, `nvcc` will occupy the vast majority of our attention. However, `ptxas` will be briefly discussed in Appendix B, which explains how to write code using Parallel Thread Execution (PTX), which is Nvidia's intermediate assembly language.

2.2.2 Library Files

The toolkit's lib folder contains several libraries that provide useful functions. This book discusses thirteen of these libraries, and Table 2.3 lists each of them.

Table 2.3

Important Libraries

Library	Description
cublas	CUDA implementation of Basic Linear Algebra Subroutines (BLAS)
cuda	Provides access to the CUDA driver (nvcuda)
cudart	Provides the CUDA runtime
cufft	CUDA implementation of the Fast Fourier Transform (FFT)
cusolver	Solves dense matrix systems of equations
cusparse	Solves sparse matrix systems of equations
nppial	Arithmetic/logical operations
nppicc	Color conversion and sampling
nppicom	JPEG compression and decompression
nppidei	Data exchange and initialization
nppif	Computer vision and filtering
nppig	Geometry transformation functions
nppim	Morphological operations

Later chapters will explore many of these libraries and their features. For example, Chapter 6 will explore the cuBLAS library, Chapter 7 will introduce the cuSOLVER library, and Chapter 8 will present the cuFFT library.

2.2.3 Header Files

To access functions in the library files, source files need to include the appropriate header files (*.h). Most of these files have the same names as the corresponding library files. That is, the functions in the cublas library are declared in cublas.h, the functions in the cusolver library are declared in cusolver.h, and so on.

The toolkit's include folder contains a vast number of header files, but for most CUDA programs, two files are particularly important:

1. cuda.h — declares low-level functions that access the device driver
2. cuda_runtime.h — declares functions that access the CUDA runtime

The relationship between the CUDA runtime functions and the CUDA driver functions is fascinating. But a proper discussion of the CUDA Driver API will have to wait until Appendix A.

2.3 Building Applications

Many sources refer to `nvcc` as the "Nvidia compiler" or the "CUDA compiler", but these terms really aren't accurate. Nvidia's documentation refers to `nvcc` as the *CUDA compiler driver*. It's responsible for managing the build process, but it doesn't compile code by itself.

Most CUDA developers rely solely on `nvcc` during the build process, and because of its importance, a large part of this chapter is devoted to explaining its use. But the goal of this section is to introduce the process needed to build CUDA code. To be specific, the goal is to explain how nvcc transforms a source file containing C++ and CUDA into an executable.

2.3.1 Simple Project

Before I discuss `nvcc` in depth, it's a good idea to walk through an example of how it can be used to build a project. If you look through this book's example code, you'll find a project named ch02_intro, which contains a single source file named main.cu.

CUDA source files have a *.cu suffix, and main.cu is very simple. It calls on the GPU to multiply a number by 2, and Listing 2.1 shows what the code looks like.

Listing 2.1 ch02_intro/main.cu

```
// The kernel function to be executed on the GPU
__global__ void timesTwo(const float *num, float* twice_num) {
  twice_num[0] = num[0] * 2.0;
}

int main() {

  // Declare variables
  float *input, *output;
  float inputVal = 7.0, outputVal;

  // Allocate buffer for input value
  cudaError_t cudaStatus = cudaMalloc(
    reinterpret_cast<void**>(&input), sizeof(float));
  if (cudaStatus != cudaSuccess) { ... }

  // Allocate buffer for output value
  cudaStatus = cudaMalloc(
    reinterpret_cast<void**>(&output), sizeof(float));
  if (cudaStatus != cudaSuccess) { ... }

  // Set input value to 7.0
  cudaStatus = cudaMemcpy(
    input, &inputVal, sizeof(float), cudaMemcpyHostToDevice);
  if (cudaStatus != cudaSuccess) { ... }

  // Multiply input by 2
  timesTwo <<<1, 1>>> (input, output);
  if (cudaStatus != cudaSuccess) { ... }

 // Access processed output
  cudaStatus = cudaMemcpy(&outputVal, output, sizeof(float),
    cudaMemcpyDeviceToHost);
  if (cudaStatus != cudaSuccess) { ... }

 // Display result
  std::cout << inputVal << " times two is " <<
    outputVal << std::endl;
  return 0;
}
```

Chapter 3 will explain what the different functions in main.cu accomplish. For now, all you have to know is that CUDA source files (*.cu) have two parts:

- **kernel code** — one or more functions written in the CUDA programming language, execute on the device
- **host code** — written in C++, run on the development system (CPU)

In Listing 2.1, the lines at the top define a kernel function named `timesTwo`. This receives a floating-point value named `num`, multiplies it by 2.0, and stores the result in a floating-point value called `twice_num`. Chapter 4 will explain the fundamentals of kernel development.

Most of the code in the listing is written in regular C++. Nvidia doesn't provide a compiler for C/C++, so you'll need to have one installed on your system. On Linux, `nvcc` will access gcc or g++. On Windows, `nvcc` will access the cl.exe compiler provided by Microsoft's Visual Studio. Throughout this book, I'll refer to the C/C++ compiler as the *host compiler.*

If you've installed the CUDA Toolkit and a host compiler is available, you can build main.cu from the command line with the following command:

```
nvcc main.cu
```

If the build succeeds, you'll find a new executable in the current directory. If you run it, it will send a value of 7.0 to the GPU, which will double the value and return a value of 14.0. Then the program will print the following statement:

```
7 times two is 14
```

Of course, this is a trivial example involving a single file. Later chapters will present more interesting applications.

2.3.2 The Build Process

Many developers run builds without understanding the sequence of steps involved. But the better you understand what's happening, the better you'll be able to handle strange errors and customize the build process for your application.

The process of building a CUDA project containing C/C++ code consists of (at least) six steps:

1. After preprocessing, the `cudafe++` utility splits the source code into device code (intended to run on the GPU) and host code (intended to run on the host).

2. The host compiler (g++, gcc, or cl.exe) compiles and assembles the host code, producing an object file.
3. The `cicc` utility receives the device code and compiles it into an intermediate assembly language called PTX (Parallel Thread Execution).
4. The PTX assembler (`ptxas`) assembles the PTX code into low-level streaming assembly (SASS). The result is a CUDA binary, or *cubin*.
5. The `fatbinary` utility packages the SASS code into a fat binary (*fatbin*), which is wrapped into an object file.
6. The linker accepts the object file from the host compiler and the object file containing CUDA code, and links them together to form an executable.

Figure 2.1 shows what this process looks like graphically. None of these utilities need to be accessed manually. The entire process is managed by `nvcc`.

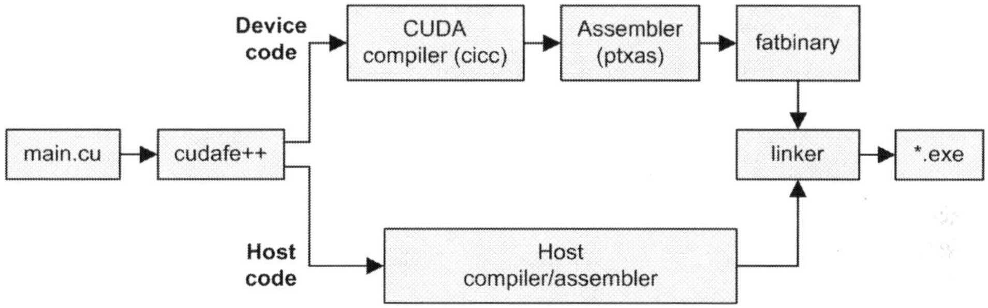

Figure 2.1 The General CUDA Build Process

For the example code in Listing 2.1, `cudafe++` will extract the following function definition and send it to `cicc` for compilation:

```
__global__ void timesTwo(const float *num, float* twice_num) {
    twice_num[0] = num[0] * 2.0;
}
```

Unlike the other utilities in the diagram, `cicc` isn't in the toolkit's bin folder. It's in the nvvm/bin folder, and it's only intended for internal use. There's no documentation on this utility or help information.

As explained in Appendix A, the CUDA Driver API makes it possible to launch the PTX assembler programmatically. Appendix B explains the fundamentals of PTX development.

2.3.3 Build Flags

Now that you have a basic grasp of the build process, you'll be able to understand the flags that tell nvcc how to perform the build. This discussion presents many, but not all of these flags. You can see the full list by running `nvcc --help` on a command line.

This discussion splits these flags into four categories:

1. **File specification** — identify files and directories used in the build
2. **Build behavior** — control how the compiler and linker operate
3. **Phase configuration** — control how a phase of the build is performed
4. **Code generation** — control the types of files generated by the build

Many of these flags have a long version that starts with two dashes and an abbreviated version that starts with one.

File Specification

If you want nvcc to use specific files in the build or generate specific files, you'll need to use flags that identify the files' names. Table 2.4 lists twelve of these flags.

Table 2.4

File Specification Flags

Short Form	Long Form	Description
-o	--output-file	The name and location of the output file
-odir	--output-directory	The directory where the output file should be stored
-l	--library	Libraries used during the linking stage
-L	--library-path	Paths containing library files
-D	--define-macro	Macro definitions used during the build
-I	--include-path	Directories containing header files
-include	--pre-include	Identify individual header files
-isystem	--system-include	The system include paths
-ccbin	--compiler-bindir	The location of the host compiler
-cudart	--cudart	The type of CUDA runtime library to be linked
-cudadevrt	--cudadevrt	The type of CUDA device runtime library to be linked

`-use-local-env`	`--use-local-env`	Use current environment variables

Most of these flags closely resemble the flags used by the GNU build tools, such as gcc and g++. For example, the following command tells `nvcc` to compile `source.cu` into an executable named `dest` and link in the `math` library located in the `mathdir` folder:

```
nvcc source.cu -o dest -l math -L mathdir
```

The last four flags are specific to CUDA development. By default, `nvcc` assumes the host compiler can be accessed through the PATH variable. But you can specify the host compiler's location by setting the `-ccbin` flag.

Most CUDA applications access the CUDA runtime library (cudart), which simplifies the process of accessing the device. The `-cudart` flag specifies how the library should be linked, and this can be set to `none`, `shared`, and `static`.

Similarly, the device runtime library (cudadevrt) is concerned with allocating device memory and launching kernels on the device. Its linking can be configured by setting `-cudadevrt` to `static` or `none`.

Build Behavior

Optimization and link-time optimization (LTO) can provide significant improvements in performance, so it's a good idea to know about the flags that configure build behavior. Table 2.5 lists twelve of these flags and provides a description of each.

Table 2.5

Build Behavior Flags

Short Form	Long Form	Description
`-O`	`--optimize`	Set optimization level for host compiler
`-dopt`	`--dopt`	Device code optimization
`-Ofc`	`--Ofast-compile`	Set fast-compile level for device
`-lto/-dlto`	`-lto/-dlto`	Link-time optimization (LTO) of device code
`-g`	`--debug`	Generate debug information for host code
`-G`	`--device-debug`	Generate debug information for device code
`-lineinfo`	`--generate-line-info`	Generates line number information in the device code

-noeh	--no-exceptions	Disable exception handling in host code
-shared	--shared	Generate shared library during linking
-x	--x	Set the language of source files
-std	--std	Set the C++ version
-m	--machine	Set the device architecture

The first flag, `-O`, sets the optimization level for the host code only, and doesn't affect the device code. For gcc and g++, the optimization level can be set to 0 to 3. For cl.exe, it can be set to 1 (small code) or 2 (fast code). For a full description of optimization, check the documentation of your host compiler.

In contrast, the device optimization flag (`-dopt`) doesn't have levels. The only allowed value is `on`. This is turned on by default, but if the debug flag is set (`-G`), device optimization won't be used.

The fast compile optimization flag (`-Ofc`) tells the compiler whether to prioritize compilation speed or performance during the build. It can be set to one of four values:

- `0` — Fast compilation is disabled (default)
- `min` — Minimal impact on compile time and runtime
- `mid` — Balance compile time and runtime, disabling expensive optimizations
- `max` — Focus on compile time, disabling many optimizations

Normal optimization is performed by examining individual source files during compilation. But if link-time optimization (LTO) is enabled, the linker will perform optimization by analyzing multiple source files at once. For `nvcc`, the `-lto` and `-dlto` flags are used for LTO on the device code, not the host code. It's important to follow the flag with `-dlink`, which tells the linker to link object files with relocatable device code.

To step through an executable, a debugger needs to access symbols in the executable's content. With `nvcc`, symbols are embedded in the host code if the `-g` flag is set. Debugging symbols are embedded in the device code if the `-G` flag is set. If the `-lineinfo` flag is set, line number information will be generated.

`nvcc` determines which language or languages to compile by looking at the suffixes of the input files. The input file language can be configured by setting the `-x` flag to `c`, `c++`, or `cu`. For C++ compilation, the target standard can be configured with the `-std` flag, which can be set to `c++03`, `c++11`, `c++14`, `c++17`, or `c++20`.

The last flag, `--machine`, specifies the architecture of the target device. The default value is `64`, which is the only accepted value. This can also be specified by using the `-m64` flag.

Phase Configuration

By default, `nvcc` compiles all of the input files and links them into an executable that can run on the host system. But with the appropriate flags, you can customize which phases of the build are performed. Table 2.6 lists eleven of these flags and provides a description of each.

Table 2.6

Phase Configuration Flags

Short Form	Long Form	Description
`-E`	`--preprocess`	Preprocess all input files
`-c`	`--compile`	Compile each input file into an object file, but don't link
`-dc`	`--device-c`	Compile input files into an object file that contains relocatable device code
`-dw`	`--device-w`	Compile input files into an object file that contains executable device code
`-cuda`	`--cuda`	Compile *.cu files to .cu.cpp.ii output
`-ptx`	`--ptx`	Compile *.cu files to *.ptx files, discarding host code
`-cubin`	`--cubin`	Compile *.cu files to *.cubin files, discarding host code
`-fatbin`	`--fatbin`	Compile *.cu files to *.fatbin files, discarding host code
`-link`	`--link`	Compile and link all input files
`-dlink`	`--device-link`	Link object files with relocatable device code and device files into an object file with executable device code
`-lib`	`--lib`	Compile input files into object files and link them to the output library file

Most of the build phases, like preprocessing and compiling, are performed by default. However, the `-c` flag specifies that each *.cu file should be compiled into an object file, but not linked.

By default, the goal of the build process is to produce an executable. But `nvcc` will produce a *.ptx file containing Parallel Thread Execution (PTX) code if the `-ptx` flag is given. It will produce a CUDA binary, *.cubin, containing Streaming Assembly (SASS) code if `-cubin` is given.

If the `-fatbin` flag is set, `nvcc` will generate a fatbinary file, *.fatbin. This contains multiple versions of the same GPU code compiled for different device architectures. This allows the same code to run on current and future hardware.

Code Generation

If you run `nvcc main.cu` without any flags, the code in the executable will target the earliest architecture supported by your version of the CUDA Toolkit. This ensures that your code will run on as many different devices as possible. For example, my version of the CUDA Toolkit is 13.1, and when I run a command like `nvcc main.cu`, the resulting code targets Turing devices and later.

You can exert control over the code generated by the build process using the flags listed in Table 2.7.

Table 2.7

Code Generation Flags

Short Form	Long Form	Description
`-arch`	`--gpu-architecture`	The class of the target device architecture
`-code`	`--gpu-code`	The name of the target hardware to assemble and optimize PTX
`-rdc`	`--relocatable -device-code`	Enable or disable generation of relocatable device code (needed for dynamic parallelism)
`-e`	`--entries`	The entry functions for which code must be generated
`-use_ fast_math`	`--use_fast_math`	Use the fast math library
`-ftz`	`--ftz`	Specifies if denormal values should be set to zero
`-fmad`	`--fmad`	Enables/disables the conversion of floating-point multiplies and adds into multiply-add (MAD) operations

The first two options are important but confusing, because they accomplish very similar tasks. The `-arch` flag specifies the version of PTX that should be generated during compilation. This identifies a set of features, and must be set to a value with the form `compute_`*N*, where *N* is a number like `60`, `70`, or `80`.

The `-code` flag specifies the version of the machine code that should be inserted into the final binary. This is usually set to a value of the form `sm_`*N*, where *N* is a number like `60`, `70`, or `80`.

In the values `compute_`*N* and `sm_`*N*, the number *N* is called the *compute capability*. This identifies the unique set of features for an architecture, and the higher the number, the more advanced the architecture. Table 2.8 lists compute capability values from 50 to 100.

Table 2.8
Compute Capability Values

Value	Architecture	Example Hardware
50	Maxwell	GTX 980
60	Pascal	GTX 1080, Tesla P100
70	Volta, Tesla	V100
75	Turing	RTX 2080, T4
80	Ampere	A100
86	Ampere	RTX 3080, RTX 3090
89	Ada Lovelace	RTX 4090, L40
90	Hopper	GH200, H200, H100
100	Blackwell	GB200, B200, B300

When calling `nvcc`, it's a good idea to set `-arch` to the lowest compute capability that will meet the application's requirements. The `-code` flag should be set to the exact compute capability of the target device.

An example will clarify how these flags work. The following command compiles main.cu so that the compiled PTX can target any architecture after Turing. However, the generated executable targets an Ampere device like the RTX 3090:

```
nvcc main.cu -arch=compute_75 -code=sm_86
```

In this example, the executable won't run on any architecture before `sm_86`. However, `nvcc` will embed PTX in the executable if the `-code` has an associated `compute_`*N* flag. For example, the following command compiles a binary for Ampere but embeds PTX that can be run on any device after the Turing architecture:

```
nvcc main.cu -arch=compute_75 -code=sm_86,compute_75
```

If the driver can't find binary code for the connected GPU, it will look for embedded PTX in the executable. If it finds PTX that can run on the target device, the driver will automatically assemble the code into SASS and deploy the binary to the hardware.

This is called *just-in-time (JIT) compilation*, and this makes it possible for existing code to run on future hardware. For this reason, it's a good idea to embed PTX in any executables that you send to customers.

2.4 Building the Example Code

If you want to access the example code discussed in this book, open a browser and visit one of two sites, depending on your operating system:

- https://github.com/mattscar/cuda-book-win — provides the example code in a Visual Studio solution
- https://github.com/mattscar/cuda-book-gnu — provides the example code in a CMake project

In packaging the examples, I've done my best to make the build process as simple as possible. The following discussion explains how to build the code on Windows using Visual Studio and how to build the code on Linux systems using CMake and GNU tools.

2.4.1 Building on Windows

For Windows users, the example code is provided in a Visual Studio solution containing twenty-nine projects. Figure 2.2 shows what this looks like on my system:

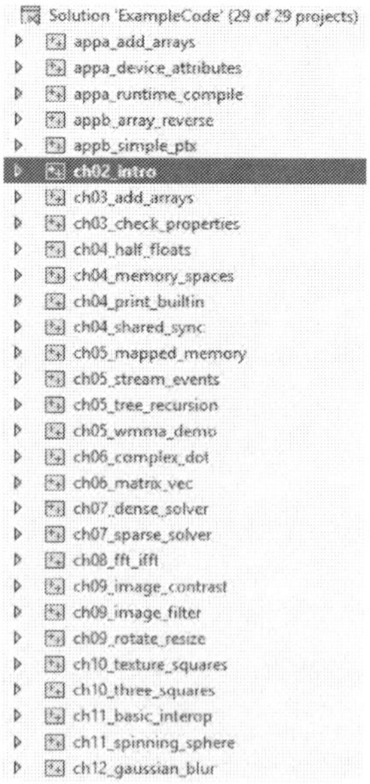

Figure 2.2 Projects in the Example Solution

It's important to understand how these projects access the CUDA Toolkit. Each has a file named <proj_name>.vcxproj that defines the project's properties. Each file reads the CUDA_HOME variable and creates variables inside a property group:

```
<PropertyGroup>
  <CudaPathClean Condition="'$(CUDA_PATH)' != ''">
    $(CUDA_PATH.TrimEnd('\'))
  </CudaPathClean>
  <CudaVersionDir Condition="'$(CudaPathClean)' != ''">
    $([System.IO.Path]::GetFileName($(CudaPathClean)))
  </CudaVersionDir>
  <DetectedCudaVer Condition="'$(CudaVersionDir)' != ''">
    $(CudaVersionDir.Replace('v', ''))
  </DetectedCudaVer>
</PropertyGroup>
```

After defining these properties, the project file can access the properties of the CUDA Toolkit with the following import group definition:

```
<ImportGroup Label="ExtensionSettings">
  <Import Project="$(VCTargetsPath)\BuildCustomizations\
    CUDA $(DetectedCudaVer).props" Condition="'$(DetectedCudaVer)'
    != '' And Exists('$(VCTargetsPath)\BuildCustomizations\CUDA
    $(DetectedCudaVer).props')" />
</ImportGroup>
```

This accesses properties for the latest installed CUDA Toolkit. As a result, you can specify CUDA library names and header files without having to enter the toolkit's path.

For this book, the most important CUDA library is cudart.lib, which provides the CUDA Runtime. This is a shared library, and when an executable runs, it needs to access cudart64_13.dll in the toolkit's bin\x64 folder. If you get any errors related to this, it's important to make sure the DLL's location is part of your PATH variable.

If you right-click the solution and select **Build Solution**, Visual Studio will build most of the projects. By default, it doesn't build the projects that require cuDSS and OpenGL. cuDSS installation is explained in Chapter 7 and OpenGL installation is explained in Chapter 10.

Once these packages are installed, you can add these projects to the solution build by right-clicking the solution and selecting **Properties**. In the dialog that appears, go to the **Configuration** item and check the unchecked projects in the **Build** column.

If a project's build succeeds, the executable and required files will be stored in the solution's x64\Debug folder. If you open a command prompt and change to this directory, you'll be able to run each executable from the command line.

2.4.2 Building on Linux

The Linux example code is easy to understand and build. The top-level folder contains a file named CMakeLists.txt and a subfolder for each project. If you've installed CMake and the CUDA Toolkit, the build process consists of five steps:

1. Open a terminal and change to the top-level directory
2. Create a build folder with `mkdir build`
3. Enter the build folder with `cd build`
4. Generate the makefiles with `cmake ..`
5. Build each project with `cmake --build .`

The last step tells CMake to enter each subdirectory listed in CMakeLists.txt and build the source code into an executable. When this finishes, the build folder will contain a subfolder for each project. Each subfolder will contain the compiled executable that can be run from the command line.

An example will help make this clear. Once the ch03_add_arrays project is built, the build folder will contain a subfolder named ch03_add_arrays. If the build succeeded, the subfolder will contain an executable named ch03_add_arrays.

The last step of the process builds most of the book's projects, but not all of them. The ch07_sparse_solver project requires cuDSS to be installed and the projects in Chapters 10, 11, and 12 require OpenGL, GLUT, and GLEW. Chapter 7 explains how to install cuDSS and Chapter 10 explains how to install OpenGL, GLUT, and GLEW.

Once these additional packages have been installed, the projects can be added to the build by uncommenting lines in the top-level CMakeLists.txt. For example, once cuDSS is installed, the ch07_sparse_solver project can be added to the build by removing the pound sign (#) in the following line:

```
#add_subdirectory(ch07_sparse_solver)
```

Similarly, when OpenGL, GLUT, and GLEW are available, the `add_subdirectory` lines for the ch10_texture_squares, ch10_three_squares, ch11_basic_interop, ch11_spinning_sphere, and ch12_gaussian_blur projects can be uncommented.

In addition to installing OpenGL, GLUT, and GLEW, many Linux systems need to be specifically directed to use the Nvidia device for graphics. On my system, I set the `__NV_PRIME_RENDER_OFFLOAD` environment variable to `1`. I also set the `__GLX_VENDOR_LIBRARY_NAME` environment variable to `"nvidia"`. The second variable tells OpenGL to rely on Nvidia's OpenGL driver for graphical rendering.

2.5 Summary

This chapter has focused on the CUDA Toolkit, which provides the fundamental libraries, header files, and utilities needed to build CUDA applications. The first part of the chapter explained how to download and install the toolkit, and the rest of the chapter explored its content.

The three most important folders in the toolkit are bin, include, and lib. The bin folder contains utilities that make it possible to build and analyze CUDA applications. The most important of them is `nvcc`, which manages the build process. As discussed in this chapter, this accepts a vast number of flags that configure how the build should be performed.

The toolkit's include and lib folders provide the files needed to build C++ source code into executables. The most important library is cudart, which provides the functions that make up the CUDA Runtime. These functions perform basic CUDA operations like allocating device memory and transferring data between the CPU and GPU. Chapter 3 will explore these functions in detail.

Other libraries in the lib folder will be discussed in later chapters. Chapter 6 will explore the cuBLAS library, Chapter 7 will introduce the cuSOLVER library, and Chapter 8 will present the cuFFT library. In each case, the chapter will walk through example code that demonstrates how the library's functions can be used.

Chapter 3

Programming Fundamentals

The vast majority of CUDA programs are focused on processing numbers. These numbers may represent pixel values, signal samples, or weights in a neural network, but the procedure is generally the same: send unprocessed numbers to the device, have the device process the numbers, and then transfer the processed numbers from the device to the host.

To implement this procedure in code, an application needs to perform at least five steps:

1. Allocate blocks of memory on the target device to store the input data and processed results.
2. Transfer input data from the host to the device's allocated memory.
3. Tell the device to process the data. This may entail executing kernels or calling library functions.
4. When the processing is complete, transfer the processed results from the device's allocated memory to the host.
5. Free the allocated blocks of memory on the device.

This chapter explores each of these steps. The first part explains how to perform memory operations, such as allocating and freeing blocks of device memory. The next part shows how to transfer data between blocks of memory on the host and blocks of memory on the device.

The chapter ends with a discussion of low-level hardware access and error checking. The last example project shows how to access and display properties of a target device.

3.1 Memory Operations

Before an application can send processing tasks to a GPU, it needs to access the device's memory. Using CUDA, applications can allocate memory blocks, free memory, transfer data between the host and device, and initialize device memory. Nvidia provides a wide range of functions for memory management, but this section focuses on the ones that are most frequently used.

3.1.1 Allocating and Freeing Device Memory

A device can't process data unless the data is stored in the device's allocated memory. The simplest function for this purpose is `cudaMalloc`, which serves the same purpose as the `malloc` (memory allocation) function used in standard C. That is, it reserves a portion of memory with a given size and returns a pointer to the memory region:

```
cudaError_t cudaMalloc(void** memPtr, size_t size)
```

This function allocates a block of memory on the device that occupies the given number of bytes. There are three points to be aware of:

1. The function's name starts with `cuda`, which means it's provided by the CUDA Runtime API. If a function's name starts with `cu`, it's provided by the CUDA Driver API. Appendix A discusses the Driver API in detail.
2. The first parameter, `memPtr`, points to the variable that will hold the address of the allocated memory. Because the memory resides on the device, `memPtr` can't be used in traditional pointer operations. It's really just an identifier for the memory that can be used in other CUDA functions.
3. Like most functions in the CUDA Runtime API, this returns a `cudaError_t` that identifies the status of the operation. This takes values of the `cudaError` enumerated type and Table 3.1 lists nine values of this type that are commonly encountered.

Table 3.1

Common Error Status Values

Value	Name	Description
0	`cudaSuccess`	No errors
1	`cudaErrorInvalidValue`	Value outside the acceptable range

2	`cudaErrorMemoryAllocation`	Failed to allocate memory
3	`cudaErrorInitializationError`	Failed to initialize the CUDA runtime
4	`cudaErrorCudartUnloading`	Attempting to access CUDA runtime after driver has been unloaded
9	`cudaErrorInvalidConfiguration`	The kernel is trying to access features that aren't supported
13	`cudaErrorInvalidSymbol`	Invalid name or identifier
21	`cudaErrorInvalidMemcpyDirection`	Invalid direction for memory copy
35	`cudaErrorInsufficientDriver`	The Nvidia driver is older than the runtime library

If a program tries to allocate more device memory than the device can provide, `cudaMalloc` will return a value of `2`. A later section will present functions that provide information about the device, such as the amount of available memory.

To free memory allocated on the device, applications need to call `cudaFree`:

```
cudaError_t cudaFree(void* memPtr)
```

The first parameter of `cudaMalloc` is a `void**` and the first parameter of `cudaFree` is a `void*`. Because of this, applications commonly create a variable as a pointer and call `cudaMalloc` with a reference to the pointer. The following code shows how this can be done:

```
float* mem;
...
cudaMalloc(reinterpret_cast<void **>(&mem), 64);
...
cudaFree(mem);
```

Because `mem` is declared as a `float*`, the kernel will access the data in the allocated memory as an array of `float`s. In this example, the memory region will contain 64 bytes, so the kernel will be able to access 16 `float`s.

In addition to `cudaMalloc` and `cudaFree`, CUDA provides other functions that allocate and deallocate memory. The difference is that these functions access special types of memory or create regions of multidimensional data. Table 3.2 lists six of them, including `cudaMalloc` and `cudaFree`.

Table 3.2
Allocation/Deallocation Functions

Function	Description
`cudaMalloc(void** mem, size_t size)`	Allocates a region of contiguous memory on the device
`cudaFree(void* mem)`	Deallocates memory on the device
`cudaMallocManaged(void** mem, size_t size, uint flags)`	Allocates a block of unified memory
`cudaMallocHost(void** mem, size_t size)`	Allocates pinned memory on the host with default characteristics
`cudaHostAlloc(void** mem, size_t size, uint flags)`	Allocates pinned memory on the host with configurable characteristics
`cudaFreeHost(void* mem)`	Deallocates memory on the host

Applications should allocate memory with `cudaMallocManaged` if the host and device share the same physical memory. This shared memory is called *unified memory*, and it's a common feature of Nvidia's embedded devices, such as the Jetson boards. However, the example projects presented in this book don't access unified memory.

The `cudaMallocHost` and `cudaHostAlloc` functions allocate memory on the *host* instead of the device. To be precise, the two functions access *pinned memory*. Data in pinned memory can't be swapped to the hard disk, so it always remains in RAM. Using pinned memory can provide substantial performance improvements, and this topic will be discussed further in Chapter 5.

3.1.2 Transferring Data

Once you've allocated a block of memory on the device, there are two main ways to assign data to it:

1. Call one of the `cudaMemcpy` functions to transfer data from host memory to the device memory block.
2. Call one of the `cudaMemset` functions to initialize bytes of the device memory block to a specific value.

CUDA provides several functions that transfer data from one memory location to another. The simplest function is `cudaMemcpy`, whose signature is given as follows:

```
cudaMemcpy(void* dst, const void* src, size_t count,
  cudaMemcpyKind kind)
```

In this function, `dst` points to the destination memory to receive data, `src` points to the origin of the data to be transferred, and `count` identifies how many bytes should be transferred from `src` to `dst`. The last argument identifies the nature of the data transfer, and can take one of four values:

1. `cudaMemcpyHostToHost` — copies data from one address on the host to another
2. `cudaMemcpyHostToDevice` — copies data from a host address to a region of data on the device
3. `cudaMemcpyDeviceToHost` — copies data from a memory region on the device to a location on the host
4. `cudaMemcpyDeviceToDevice` — copies data from a memory region on the device to another memory location on the device

The following code demonstrates how `cudaMemcpy` can be used. It allocates memory on the device and then transfers five `float`s from the host to the device.

```
// Host data
const int N = 5;
float host_data[N] = {0.0f, 1.0f, 2.0f, 3.0f, 4.0f};

// Device data
float* dev_data;
cudaMalloc(reinterpret_cast<void **>(&dev_data), N*sizeof(float));

// Copy data from host to device
cudaMemcpy(dev_data, host_data, N * sizeof(float),
   cudaMemcpyHostToDevice);
```

When `cudaMemcpy` executes, the device can't perform other operations until the bytes have been initialized. But devices can perform memory operations while kernels are being executed. This is called *asynchronous operation*, and applications can take advantage of this by calling `cudaMemcpyAsync`:

```
cudaMemcpyAsync(void* dst, const void* src, size_t count,
  cudaMemcpyKind kind, cudaStream_t stream = 0)
```

Asynchronous functions like this can provide significant performance improvement. Chapter 5 will discuss asynchronous functions, streams, and other advanced topics.

3.1.3 Initializing Memory

If you just want to fill a block of memory with a value, you don't need `cudaMemcpy`. The `cudaMemset` function works like the regular `memset` function in C/C++, and its signature is given as follows:

```
cudaMemset(void* mem, int value, size_t count)
```

As shown, the function accepts three arguments:

- `mem` — pointer to the allocated memory on the device
- `value` — value used to initialize bytes of the allocated memory
- `count` — the number of bytes to be initialized with the value

`cudaMemset` is simple to use, but it's important to understand that the value is assigned to *bytes* in the buffer. For example, the following code sets the first 16 bytes in the `data` buffer to a value of `0x5`:

```
cudaMemset(data, 0x5, 16);
```

When this code executes, the device can't perform other operations until the bytes have been initialized. But applications can initialize memory and process data at the same time by calling the asynchronous version of `cudaMemset`. This function is called `cudaMemsetAsync`:

```
// Declare a stream
cudaStream_t stream;

// Create the stream
cudaStreamCreate(&stream);

// Perform asynchronous initialization
cudaMemsetAsync(data, 7, 16, stream);
```

Asynchronous operations can improve performance because they tell the device to perform memory operations and computational operations at the same time. But more code is needed because the application needs to create and access streams. Chapter 5 goes into greater detail how streams work with asynchronous operations.

3.2 Threads and Thread Blocks

To fully leverage the power of an Nvidia GPU, a program must launch vast numbers of threads, typically tens or even hundreds of thousands of threads. This section explains how to create threads in CUDA and organize them so that kernel functions run efficiently in parallel.

There are three fundamental points to understand:

1. To execute threads in parallel, a program needs to create one or more *thread blocks*. A thread block is a container of threads, and threads in the same block can access the same shared memory.
2. The maximum number of threads in a thread block depends on the device. Currently, the maximum is 1,024 threads per block.
3. If a program needs more threads than a single block can execute, it can create multiple blocks inside a *grid*. A grid can contain unlimited numbers of thread blocks, but threads in different blocks don't execute at the same time.

CUDA doesn't provide special functions that create thread blocks and grids. Instead, programs invoke kernel functions using special statements that identify the thread block size and grid size.

A full discussion of kernel coding will have to wait until Chapter 4. For now, suppose the name of the kernel function is `foo`. We want to execute `foo` with two thread blocks where each block contains sixteen threads. In this example, the grid size is `2` and the thread block size is `16`, so we can call `foo` with the following statement:

```
foo <<<2, 16>>>(...);
```

As shown, the name of the kernel function, `foo`, is followed by the grid size (number of thread blocks per grid) and the block size (number of threads per block). As a result, the GPU will execute foo with a total of 32 threads, but the threads don't necessarily execute at the same time because there are two thread blocks.

Since most devices support thread block sizes of 1,024 threads, we can execute `foo` with all the threads in a single thread block with the following statement:

```
foo <<<1, 32>>>(...);
```

Within the parentheses, the kernel function accepts one or more pointers to memory blocks that have been allocated on the device. The following code shows how this works.

```
// Host data
const int N = 3;
float host_data[N] = {0.0f, 1.0f, 2.0f};

// Allocate memory on the device
float *input, *output;
cudaMalloc(reinterpret_cast<void **>(&input), N * sizeof(float));
cudaMalloc(reinterpret_cast<void **>(&output), N * sizeof(float));

// Copy data from host to device
cudaMemcpy(output, host_data, N * sizeof(float),
   cudaMemcpyHostToDevice);

// Execute kernel with 32 threads in one block
foo <<<1, 32>>>(input, output);
```

In this case, the kernel function `foo` receives two parameters, and both are pointers to allocated memory on the device. The first pointer, `input`, represents a memory block that has been initialized with three values. The second pointer, `output`, represents a memory block whose values will be set by the kernel function.

In the kernel execution statement, `foo` is followed by `<<<1, 32>>>`. This tells the device to execute the kernel with one thread block containing 32 threads. Threads in the same block don't necessarily execute simultaneously, but they can access the same region of shared memory.

For a better understanding of threads, thread blocks, and kernel functions, you need to be familiar with CUDA's programming model, which will be discussed in Chapter 4. Right now, I'd like to walk through a simple example.

3.3 Example Project – Adding Arrays

At this point, you should understand how to access device memory and tell the device to execute a kernel with multiple threads. To demonstrate how these operations can be performed in practice, this section walks through the code in an application that adds two arrays together.

To be specific, this section looks at the ch03_add_arrays project in this book's example code. The code in main.cu performs five operations:

1. Calls `cudaMalloc` to allocate memory on the device for two input arrays and the output array.
2. Calls `cudaMemcpy` to transfer data from the host into the two input arrays.

3. Invokes the `addArrays` kernel and tells the device to execute it with six threads.
4. Transfer data from the output array from the device to the host.
5. Prints the content of the output array to standard output.

Listing 3.1 presents the content of the main.cu source file. You can build this from the command line with the command `nvcc main.cu`.

Listing 3.1 ch03_add_arrays/main.cu

```
// The kernel function to be executed on the GPU
__global__ void addArrays(const float* a, const float* b,
  float* c) {

  int i = threadIdx.x;
  c[i] = a[i] + b[i];
}

int main() {
  const int N = 6;
  float *arrayA, *arrayB, *arrayC;
  cudaError_t cudaStatus;

  // Allocate buffer for the first input array
  cudaStatus = cudaMalloc(
    reinterpret_cast<void**>(&arrayA), N * sizeof(float));
  if (cudaStatus != cudaSuccess) {
    std::cerr << "Failed to allocate first input array"
      << std::endl;
  }

  // Allocate buffer for the second input array
  cudaStatus = cudaMalloc(
    reinterpret_cast<void**>(&arrayB), N * sizeof(float));
  if (cudaStatus != cudaSuccess) {
    std::cerr << "Failed to allocate second input array"
      << std::endl;
  }

 // Allocate buffer for the output array
  cudaStatus = cudaMalloc(
    reinterpret_cast<void**>(&arrayC), N * sizeof(float));
  if (cudaStatus != cudaSuccess) {
    std::cerr << "Failed to allocate output array" << std::endl;
  }
```

Listing 3.1 ch03_add_arrays/main.cu (Continued)

```
  // Set content of first input array
  float aData[N] = { 2.0f, 4.0f, 6.0f, 8.0f, 10.0f, 12.0f };
  cudaStatus = cudaMemcpy(arrayA, aData, N * sizeof(float),
    cudaMemcpyHostToDevice);
  if (cudaStatus != cudaSuccess) {
    std::cerr << "Failed to transfer data to first input array"
      << std::endl;
  }

  // Set content of second input array
  float bData[N] = { 3.0f, 6.0f, 9.0f, 12.0f, 15.0f, 18.0f };
  cudaStatus = cudaMemcpy(arrayB, bData, N * sizeof(float),
    cudaMemcpyHostToDevice);
  if (cudaStatus != cudaSuccess) {
    std::cerr << "Failed to transfer data to second input array"
      << std::endl;
  }

  // Execute kernel
  addArrays <<<1, N>>> (arrayA, arrayB, arrayC);
  if (cudaStatus != cudaSuccess) {
    std::cerr << "Failed to execute kernel" << std::endl;
    return 1;
  }

  // Access processed output
  float cData[N];
  cudaStatus = cudaMemcpy(cData, arrayC, N * sizeof(float),
    cudaMemcpyDeviceToHost);
  if (cudaStatus != cudaSuccess) {
    std::cerr << "Failed to transfer output array to host"
      << std::endl;
  }
  cudaFree(arrayA); cudaFree(arrayB); cudaFree(arrayC);

  // Display result
  for (int i = 0; i < N; i++) {
    std::cout << aData[i] << " plus " << bData[i]
      << " equals " << cData[i] << std::endl;
  }
  return 0;
}
```

With the exception of the kernel definition at the top, most of this code should look straightforward. It calls `cudaMalloc` three times to allocate portions of device memory for the two input arrays and the output array. Then it calls `cudaMemcpy` twice to set values in the two input arrays.

Next, the source code invokes the kernel with the following statement:

```
addArrays <<<1, N>>> (arrayA, arrayB, arrayC);
```

This tells the device to execute the `addArrays` kernel with one thread block and N threads in the block. The kernel function should receive the three allocated memory blocks (`arrayA`, `arrayB`, and `arrayC`) as arguments.

After the kernel execution completes, the program transfers the content of `arrayC` to the host. Then it prints output to verify that the program has completed successfully.

3.4 Accessing Hardware

In the preceding example, the application didn't check the capabilities of the target device or select one device out of many. It simply picked the first connected device and assumed it could execute the kernel. This is the default behavior of applications based on CUDA's runtime library.

But in many cases, a program needs information about the connected hardware. For this reason, the CUDA runtime library provides functions that access the connected device or devices. Table 3.3 lists twelve of the most useful device management functions.

Table 3.3

Device Management Functions (Abridged)

Function	Description
`cudaGetDeviceCount(int* count)`	Returns the number of CUDA-compatible devices
`cudaSetDevice(int dev)`	Sets the device to be used for GPU operation
`cudaGetDevice(int* dev)`	Provides the ID of the current device
`cudaGetDeviceFlags( unsigned int* flags)`	Gets the flags of the current device
`cudaSetDeviceFlags( unsigned int flags)`	Sets the flags used for device operations

`cudaInitDevice(int dev,`
`unsigned int deviceFlags,`
`unsigned int flags)`
	Initialize current device

`cudaGetDeviceProperties(`
`cudaDeviceProp* prop,`
`int device)`
	Access properties of the given device

`cudaChooseDevice(int* dev,` `const cudaDeviceProp* prop)`	Access the device that most closely matches the given properties

`cudaDeviceGetAttribute(`
`int* val,`
`cudaDeviceAttr attr,`
`int dev)`
	Provides information for the given attribute

`cudaDeviceGetLimit(`
`size_t* pValue,`
`cudaLimit limit)`
	Provides the specified limit for the device

`cudaDeviceSetLimit(`
`cudaLimit limit,`
`size_t value)`
	Sets the given limit for the current device

`cudaDeviceReset()`	Reset device's state and free allocated memory

In many of these functions, the first parameter is a pointer. When the function returns, the pointer references the desired value. For example, `cudaGetDeviceCount` provides a pointer to the number of connected devices, and the following code shows how it can be used:

```
// Declare the device Count
int devCount = 0;

// Access the device count
cudaGetDeviceCount(&devCount);

// Display the device count
std::cout << "Number of devices: " << devCount << std::endl;
```

CUDA assigns an integer ID to each connected device. If you know the ID of the device you'd like to use, you can call `cudaSetDevice` to set the device to be used for CUDA operations. You can also access the ID of the current device by calling `cudaGetDevice`.

Every device has flags that identify its runtime behavior. These are represented as bits of an unsigned integer value. Each flag setting has a name and Table 3.4 lists the names and bit values of each flag.

Table 3.4
Device Flags

Flag	Bit Value	Description
`cudaDeviceScheduleAuto`	`0x00`	Default scheduling
`cudaDeviceScheduleSpin`	`0x01`	CPU spins while waiting for device
`cudaDeviceScheduleYield`	`0x02`	CPU yields while waiting for device
`cudaDeviceScheduleBlockingSync`	`0x04`	CPU sleeps while GPU operations
`cudaDeviceMapHost`	`0x08`	Enables mapping host memory into device memory
`cudaDeviceLmemResizeToMax`	`0x10`	Rounds memory allocations to maximum size

The first four flags identify how the host should behave while the device performs its operations, and only one of their bits can be set high. If the 0x1 bit is high, the host will repeatedly check to see if the device operation has finished. This is called *spinning*, and it guarantees that the host will respond promptly.

If the 0x2 bit is high, the host will be allowed to process other threads while the device works, and if the 0x4 bit is high the host will block until it receives a signal from the device. By default, the host will spin if the number of processors is greater than or equal to the number of CUDA devices, and yield to other threads otherwise.

Chapter 5 will discuss the advantages of using pinned memory on the host, also called mapped memory. The `cudaDeviceMapHost` flag must be set before this type of memory can be allocated.

To demonstrate how flags can be set, the following line of code enables mapped memory and specifies that the host should spin while waiting for the device:

```
cudaSetDeviceFlags(cudaDeviceScheduleSpin | cudaDeviceMapHost);
```

To understand `cudaSetDevice` and `cudaInitDevice`, you need to be familiar with CUDA *contexts*. A context holds all the data that the device needs to perform its work. This includes memory allocations, compiled kernel code, and other state information. A device can't execute until a context has been initialized.

The `cudaSetDevice` function associates the host thread with the device and creates and initializes a context for the device. When a program calls `cudaInitDevice`, it creates and initializes the context with the given flags, but it doesn't associate the host thread with the device. In essence, this program prepares a context for the device so that it can be selected later.

A program can access a device's properties with the `cudaGetDeviceProperties` function. The properties are provided in a `cudaDeviceProp` data structure, which has a vast number of fields. Table 3.5 lists twenty of these properties and the middle column identifies the corresponding data type.

Table 3.5

Device Properties (Abridged)

Property	Data Type	Description
`name`	`char[256]`	The device's name
`major`	`int`	The device's major version
`minor`	`int`	The device's minor version
`regsPerBlock`	`int`	Number of registers per thread block
`regsPerMultiprocessor`	`int`	Number of registers per multiprocessor
`maxThreadsDim`	`int[3]`	Maximum number of threads in each dimension
`maxGridSize`	`int[3]`	Maximum number of thread blocks in a grid for each dimension
`maxThreadsPerBlock`	`int`	Maximum number of threads in a thread block
`maxBlocksPerMultiprocessor`	`int`	Maximum number of thread blocks per multiprocessor
`maxThreadsPerMultiprocessor`	`int`	Maximum number of threads per multiprocessor
`multiProcessorCount`	`int`	Number of multiprocessors
`L2CacheSize`	`int`	Size of Level 2 cache
`totalGlobalMem`	`size_t`	Total amount of global memory
`totalConstMem`	`size_t`	Total amount of constant memory
`sharedMemPerBlock`	`size_t`	Amount of shared memory available for each thread block
`sharedMemPerMultiprocessor`	`size_t`	Amount of shared memory for each multiprocessor
`memoryPoolsSupported`	`int`	Identifies if memory pools are supported

hostNativeAtomicSupported	int	Identifies if atomic operations are supported
memoryBusWidth	int	The width of the device's memory bus
canMapHostMemory	int	Identifies if host memory can be mapped to the device

Most of these properties identify the amount of resources on the device. Programs frequently check the maxThreadsPerBlock property so that they create thread blocks with the maximum number of threads. Other popular properties are L2CacheSize, totalGlobalMem, and totalConstMem, which identify how much memory is available on the device.

The major and minor properties are particularly useful because they identify the GPU's *compute capability* (CC). Programs frequently check these properties to check the device's architecture and capabilities. You can see the full list of current CC designations by visiting https://developer.nvidia.com/cuda-gpus.

The following code demonstrates how a program can access the identifier of the current device and its properties.

```
int dev;
cudaDeviceProp prop;

// Get the ID of the current device
cudaGetDevice(&dev);

// Get the device properties of the current device
cudaGetDeviceProperties(&prop, dev);

// Display the maximum number of threads per block
std::cout << prop.maxThreadsPerBlock << std::endl;
```

In addition to accessing properties, programs can access a device's attributes. Attributes provide the same information as properties, but they're easier to access. This is because the data isn't contained in a large, single structure. Instead, the cudaDeviceGetAttribute function accepts a device ID and an attribute name and returns the corresponding value.

Many attribute names resemble corresponding property names, but they all start with cudaDevAttr. For example, the attribute corresponding to the maxThreadsPerBlock property is cudaDevAttrMaxThreadsPerBlock. The attributes corresponding to the minor and major properties are cudaDevAttrComputeCapabilityMinor and cudaDevAttrComputeCapabilityMajor.

The following code demonstrates how attributes can be accessed. It obtains the ID of the current device and then calls `cudaDeviceGetAttribute` twice to obtain the device's compute capability minor/major values:

```
int dev, minor, major;

// Get the ID of the current device
cudaGetDevice(&dev);

// Get the CC major value
cudaDeviceGetAttribute(&minor,
  cudaDevAttrComputeCapabilityMajor, dev);

// Get the CC minor value
cudaDeviceGetAttribute(&major,
  cudaDevAttrComputeCapabilityMinor, dev);

// Display results
std::cout << "Compute capability major: " << major << std::endl;
std::cout << "Compute capability minor: " << minor << std::endl;
```

Like properties and attributes, CUDA's limits provide information about the device's capabilities. What makes limits different is that they're not fixed. Programs can access a device's limit values with `cudaDeviceGetLimit` and then change limit values with `cudaDeviceSetLimit`.

Like attributes, limits are identified using values of an enumerated type. Table 3.6 lists the different identifiers and the corresponding device limit.

Table 3.6

Device Limit Identifiers

Limit ID	Description
`cudaLimitStackSize`	Device thread stack size
`cudaLimitPrintfFifoSize`	Device printf FIFO size
`cudaLimitMallocHeapSize`	Device malloc heap size
`cudaLimitDevRuntimeSyncDepth`	Device runtime synchronize depth
`cudaLimitDevRuntimePendingLaunchCount`	Device runtime pending launch count
`cudaLimitMaxL2FetchGranularity`	Maximum fetch granularity of the device's L2 cache
`cudaLimitPersistingL2CacheSize`	Size of persisting lines in L2 cache

If a program needs to conserve memory, it can call `cudaDeviceSetLimit` to set a limit for the amount of memory that can be allocated from the heap. For example, the following code sets the maximum heap allocation size to 16 MB.

```
size_t heapSize = 16 * 1024 * 1024;
cudaDeviceSetLimit(cudaLimitMallocHeapSize, heapSize);
```

When a program is done using a device, it can call `cudaDeviceReset` to free resources associated with it, such as the context. When calling this function, it's important to make sure that there are no other host threads accessing the device.

3.5 Error Checking

As mentioned earlier, most functions provided by the CUDA runtime library return a `cudaError_t` that can be used to check the status of the function's operation. Many programs simply check to see if the returned value indicates success, and print a message if it doesn't.

CUDA provides four functions that assist with error checking. Table 3.7 lists them and provides a description of each.

Table 3.7

Error Handling Functions

Function	Description
`cudaGetErrorName(cudaError_t err)`	Returns a string containing the error's name
`cudaGetErrorString(cudaError_t err)`	Returns a string containing a description of the error
`cudaGetLastError()`	Returns the last error from a runtime call, resets error state
`cudaPeekAtLastError()`	Returns the last error from a runtime call, doesn't reset error state

The CUDA runtime keeps track of an error state, and by default, the state equals `cudaSuccess`, or 0. When a runtime function executes, the error state is set equal to the `cudaError_t` returned by the function.

When a program calls `cudaGetLastError`, the return value is the error state of the CUDA runtime. Then the error state is reset to the default of 0. But if a program calls `cudaPeekAtLastError`, the return value will be set to the runtime's error state, but the error state won't be reset.

In Nvidia's sample code, error checking is performed by surrounding runtime functions with a macro called `checkCudaErrors`. This receives a `cudaError_t` and prints a description of the error to standard output. I've written code that accomplishes the same result:

```
inline void checkErr(cudaError_t err, const char* file, int line)
{
  if (err != cudaSuccess) {
    std::cerr << "CUDA error in " << file << " at line "
      << line << ": " << cudaGetErrorString(err) << std::endl;
    std::exit(EXIT_FAILURE);
  }
}
#define checkCudaErrors(err) (checkErr((err), __FILE__, __LINE__))
```

If an error occurs, this macro prints the error's file name, line number, and description, and then exits. The C++ `iostream` header needs to be included to ensure that the `std::cerr` can be accessed.

3.6 Example Project – Device Properties

The ch03_check_properties project demonstrates how an application can select a device based on its properties. To be specific, the code in main.cu performs the following operations:

1. It determines how many CUDA-compatible devices are connected by calling `cudaGetDeviceCount`.
2. For each device, it obtains the device properties by calling `cudaGetDeviceProperties`.
3. It displays the name, major, and minor properties of each device, and then checks the maximum number of threads per block.
4. Once it determines which device provides the most threads per block, it calls `cudaSetDevice` to create the association between the host thread and the device and initialize a context for the device.

Listing 3.2 presents the code in the ch03_check_properties/main.cu source file. For each call to the runtime library, the resulting error code is checked with the `checkCudaErrors` macro discussed earlier.

Listing 3.2 ch03_check_properties/main.cu

```
int main() {

  int devCount = 0, maxThreads = -1, maxIndex = -1;
  cudaDeviceProp prop;

  // Determine the number of connected compatible devices
  checkCudaErrors(cudaGetDeviceCount(&devCount));

  // Display the number of connected devices
  std::cout << "Number of connected devices" << devCount
    << std::endl;

  // Iterate through devices
  for (int i = 0; i < devCount; i++) {

    // Get properties of device
    checkCudaErrors(cudaGetDeviceProperties(&prop, i));

    // Display name, major, minor properties:
    std::cout << "Device " << prop.name << ": "
      << prop.major << "." << prop.minor << std::endl;

    // Check maximum threads per block
    if (prop.maxThreadsPerBlock > maxThreads) {
      maxThreads = prop.maxThreadsPerBlock;
      maxIndex = i;
    }
  }

  // Set the current device to the one with
  // the most threads per block
  checkCudaErrors(cudaSetDevice(maxIndex));

  return 0;
}
```

This code is straightforward to understand. The same results can be obtained by accessing device attributes with `cudaDeviceGetAttribute`.

3.7 Summary

This chapter has presented many of the basic functions provided in the CUDA Runtime API. Without question, the most important functions to know are those that perform memory operations. No matter whether you write your own kernels or use the toolkit's libraries, you'll end up using `cudaMalloc` to allocate device memory, `cudaMemcpy` to transfer data between the host and device, and `cudaFree` to free a block of device memory.

Applications with custom kernels need to invoke the kernel function using a special function of the form `foo<<<X, Y>>>(...)`, where `X` is the number of thread blocks in the grid and `Y` is the size of each thread block. This notation may look weird now, but it will get even weirder in later chapters, which discuss shared memory and streams.

The final parts of this chapter discussed hardware access and error checking. Accessing hardware is a major concern for applications that need to know device-specific properties like the maximum number of threads and the amount of available memory. The error checking functions make it possible to display an error message and the line of code that caused the error, and they'll be used in the example code throughout this book.

Chapter 4

Kernel Development

Kernel coding is the most difficult topic presented in this book, and there are three reasons for this:

1. The kernel programming language resembles C, but has many CUDA-specific quirks and details.
2. There are several new data types and CUDA-specific memory spaces where variables can be stored.
3. To write a kernel that can take full advantage of the target device, you need to have a solid understanding of the GPU's processing model.

This chapter discusses each of these aspects of kernel development. The first part discusses the basics of kernel coding, including the new data types, memory spaces, and threads. The second part discusses GPU architecture, and explains how an Nvidia GPU can be thought of as a container of streaming multiprocessors (SMs). Each SM has a unique block of shared memory and can execute one or more thread blocks.

To ensure kernels execute quickly, you need to grasp the big picture of how threads access data in the different memory spaces. For example, global memory is necessary to store results for the host to access. But intermediate data should be stored in shared memory, where threads in the same thread block can access it. If input data never changes, it should be stored in constant memory.

The last part of the chapter discusses the important topic of synchronization. When thousands of threads execute at the same time, it's important to make sure they're all processing current data. For this reason, CUDA provides mechanisms to ensure that threads can only continue execution when every other thread has finished prior work.

4.1 Understanding Kernels

In a CUDA application, custom routines intended to be executed on a GPU are called *kernels*. If you're familiar with how functions are defined in the C programming language, the format of a kernel won't present too much difficulty.

The overall structure of a kernel function that can be launched from the host is given as follows:

```
__global__ void func_name(params) { ... }
```

The `__global__` modifier specifies that the kernel can be launched by the host or by other kernels on the device (using *dynamic parallelism*, discussed in Chapter 5). If a kernel is only intended to be executed by other kernels, the `__global__` modifier should be replaced with `__device__`.

To explain kernel coding, this presentation has three parts. First, I'll describe the parameters that can be passed into a kernel and the possible data types. Then I'll explain how variables can be declared in a kernel and how to specify a memory space for each variable. The last part of this section discusses the built-in variables that can be accessed in a kernel and the important functions they serve.

4.1.1 Parameters and Data Types

Kernel functions always return `void`, which means that input data structures and output data structures must be provided as parameters. Three types of data can be passed into a kernel:

1. Pointers to blocks of device memory
2. Pointers to blocks of pinned host memory
3. Scalars, vector types, and structures containing scalars and vector types

For parameters of the first type, Chapter 3 explained how applications can allocate blocks of device memory by calling `cudaMalloc`. This provides a pointer to the device memory that can be passed to a kernel function as an argument.

For parameters of the second type, Chapter 5 discusses the fascinating topic of pinned memory. It will also present the functions that enable and allocate pinned memory on the host.

Parameters of the third type contain primitive data, which excludes pointers. All the basic types of C/C++ are supported, including `bool`s, `char`s, `short`s, `int`s, `long`s, `long long`s, and the unsigned versions of integer types. For floating-point values, parameters can be created from `double`s, `float`s, and `half` values.

The Half Type

The `half` type isn't supported by C or C++, but the cuda_fp16.h header provides a type called `__half`, which represents half-precision floating-point values. If a floating-point value is cast to a `__half`, it can be passed to a kernel as a `__half`.

The ch04_half_floats project demonstrates how the `__half` type can be used. The code in main.cu performs five operations:

1. Initializes a `__half` variable and a `float` variable to serve as inputs.
2. Calls `cudaMalloc` to allocate device memory for the output `__half` variable.
3. Executes the `processHalf` kernel with one thread and passes the `__half` value, the `float` value, and the output memory block.
4. Calls `cudaMemcpy` to access the `__half` value in the output memory block.
5. Prints the `__half` value to standard output.

Listing 4.1 presents the code in the project's main.cu source file. Note that the kernel accepts three parameters: two primitive values (a `__half` and a `float`) and one pointer to a device memory block.

Listing 4.1 ch04_half_floats/main.cu

```
// Kernel function
__global__ void processHalf(const __half a, const float b,
  __half* c) {

  c[0] = a * (__half)b;
}

int main() {

  __half halfVal = (__half)7.0f;
  float floatVal = 3.0f;

  // Allocate buffer for the output half array
  __half *outputMem;
```

Listing 4.1 ch04_half_floats/main.cu (Continued)

```
    checkCudaErrors(cudaMalloc(
      reinterpret_cast<void**>(&outputMem), sizeof(__half)));

    // Execute kernel
    processHalf <<<1, 1>>> (halfVal, floatVal, outputMem);

    // Access processed output
    __half result;
    checkCudaErrors(cudaMemcpy(&result, outputMem,
      sizeof(__half), cudaMemcpyDeviceToHost));

    // Display results
    std::cout << "Output: " << (float)result << std::endl;

    // Free memory blocks
    cudaFree(outputMem);
    return 0;
}
```

As shown, the host passes two primitive values (`halfVal` and `floatVal`) directly to the kernel. The value of `halfVal` is set by casting a `float` to `__half`. At the end of the code, the `__half` result must be cast to a `float` before its value can be printed to standard output.

Using half-precision floating-point values instead of single-precision values can provide more efficient storage and faster data transfer. But this comes with disadvantages including lower precision and the potential for numerical instability.

Vector Types

In many cases, applications may want to access values in groups of two, three, or four, such as when processing coordinates in three dimensions. For this reason, CUDA provides *vector types*. Like an array, a vector contains multiple elements of the same type. But there are three major differences between vectors and arrays:

1. A vector can only contain a specific number of elements.
2. On the host, vectors must be created using special CUDA functions.
3. Vector elements are accessed by the letters `x`, `y`, `z`, and `w`.

Each vector type is named after a scalar type followed by the number of scalar values contained in the vector. Table 4.1 lists each of the vector types supported by CUDA.

Table 4.1

Vector Data Types

Data Type	Description
char1/char2/char3/char4	Vectors containing one or more signed 8-bit integers
uchar1/uchar2/uchar3/uchar4	Vectors containing one or more unsigned 8-bit integers
short1/short2/short3/short4	Vectors containing one or more signed 16-bit integers
ushort1/ushort2/ushort3/ushort4	Vectors containing one or more unsigned 16-bit integers
int1/int2/int3/int4	Vectors containing one or more signed 32-bit integers
uint1/uint2/uint3/uint4	Vectors containing one or more unsigned 32-bit integers
long1/long2/long3/long4_16a/long4_32a	Vectors containing one or more signed long integers
ulong1/ulong2/ulong3/ulong4_16a/ulong4_32a	Vectors containing one or more unsigned long integers
longlong1/longlong2/longlong3/longlong4_16a/longlong4_32a	Vectors containing one or more signed 64-bit integers
ulonglong1/ulonglong2/ulonglong3/ulonglong4_16a/ulonglong4_32a	Vectors containing one or more unsigned 64-bit integers
float1/float2/float3/float4	Vectors containing one or more 32-bit floating-point values
double1/double2/double3/double4_16a/double4_32a	Vectors containing one or more 64-bit floating-point values

If you look at the rows for the `longN`, `ulongN`, `longlongN`, `ulonglongN`, and `doubleN` types, you'll see that the last two types end in `_16a` and `_32a`. For example, the last two `longN` types are `long4_16a` and `long4_32a`. The `a` stands for alignment, and if a type ends with `_16a`, it will always be stored on a 16-byte boundary. If a type ends with `_32a`, it will always be stored on a 32-byte boundary.

A host application can pass vectors directly to a kernel just as it can pass `int`s, `float`s, and other scalars. But to create a vector on the host, you need to call one of the creation functions in cuda_runtime.h. These functions all have the same basic format:

```
make_type(element1, element2, element3, ...)
```

Here, *`type`* identifies the type of the vector to be created. The function's arguments will become the elements of the vector returned by the function. For example, the following code creates a `float4` vector containing the values 1.0, 2.0, 3.0, and 4.0:

```
float4 floatVec = make_float4(1.0f, 2.0f, 3.0f, 4.0f);
```

If you want to fill the elements of a vector with a single value, call the function with a single argument. To demonstrate, the following code creates a four-element vector whose elements are all set to `5`.

```
int4 intVec = make_int4(5);
```

One difference between vectors and arrays is that elements in a vector are accessed with subscripts `x`, `y`, `z`, and `w`. After the preceding example, the first element of `intVec` can be accessed as `intVec.x`, the second element as `intVec.y`, the third element as `intVec.z`, and the fourth element can be accessed as `intVec.w`.

Unlike other toolsets, CUDA does not provide special operations for vectors. That is, if you want to add vectors `a` and `b`, you can't just set `c` equal to `a + b`. Instead, you need to write code like the following:

```
float4 c =
  make_float4(a.x + b.x, a.y + b.y, a.z + b.z, a.w + b.w);
```

Because CUDA doesn't provide vector operations, using vectors doesn't provide a significant advantage over using arrays. It's a good idea to use vectors when the subscripts `x`, `y`, `z`, and `w` are more convenient to use than regular indexing.

4.1.2 Variables and Memory Spaces

Now that you understand how to provide parameters to a kernel function, you're ready to start writing the function's code. This discussion focuses on declaring variables inside the kernel, which is more complicated than declaring variables inside regular functions.

Inside a kernel, a variable can be declared as one of the following types:

1. Built-in C/C++ types; `bool`, `short`, `int`, `long long`, `float`, `double`
2. CUDA vector types, such as `int3` and `float4`
3. Structures and classes containing primitive data
4. Pointers to device memory

Memory Spaces

As a later section will explain in detail, Nvidia devices have four different types of memory that applications can access:

1. **Global memory** — accessible to every thread, large but slow
2. **Constant memory** — accessible to every thread, smaller than global memory but faster, can't be modified
3. **Shared memory** — accessible to threads inside the same thread block, smaller than global memory but faster
4. **Registers** — can only be accessed by one thread, smallest and fastest

Every thread has a small block of register memory that only it can access. When a primitive value, such as an `int` or `float3`, is declared in a kernel, it's stored by default in the registers of the thread executing the kernel, and no other thread can access it.

For example, suppose a CUDA kernel contains the following line of code:

```
int x = 5;
```

This declaration has no modifiers, so `x` will be stored in the thread's register memory. If the kernel is executed by multiple threads, each thread will have its own distinct `x` variable in its register block.

Register memory is the fastest but smallest of the four memory spaces, and many devices only provide a few kilobytes of this memory to each thread. If a kernel declares a variable that's too large to fit in register memory, the compiler stores it in *local memory*, which is a portion of global memory dedicated to per-thread data. Local memory is just as slow as global memory, so it's a good idea to store large data elsewhere.

Shared memory is faster than global memory, and the size of shared memory is much larger than the size of register memory. If a variable's declaration starts with `__shared__`, its data will be allocated from shared memory instead of register memory. The following code creates an array of 1000 `float`s in shared memory:

```
__shared__ float data[1000];
```

Every thread block has a dedicated block of shared memory, and each thread in the thread block can access the shared memory block. This makes shared memory ideal for storing intermediate results produced by computation. If threads need to access data quickly and the data is too large to be stored in register memory, it's probably a good idea to allocate the data in shared memory.

In addition to the `__shared__` specifier, CUDA provides the `__device__` specifier for variables in global memory and the `__constant__` specifier for variables in constant memory. But there's one major difference: the `__device__` and `__constant__` specifiers can only be used outside of the kernel. That is, you can't allocate new blocks of global or constant memory inside a kernel.

For example, this statement declares `x` as an array of 1000 `float`s in global memory.

```
__device__ float x[1000];
```

This statement must be placed outside of any kernel function. It handles its own memory allocation, so there's no need to call `cudaMalloc`. Once a variable has been declared with this statement, it can be used in any kernel associated with the current device. But like all data in global memory, it can be slow to access unless it's been cached.

Nvidia devices have a special cache for read-only data. It's called *constant memory*, and it's very small (64 kB). But it's much faster than global memory because multiple threads can access the cache without consistency issues. If a variable declaration is preceded by `__constant__`, the compiler will store it in constant memory. The following code provides an example.

```
__constant__ float x[1000];
```

If a variable is stored in constant memory, its value can be read inside a kernel but its value can't be changed. The host application can initialize variables in constant or global memory by calling the `cudaMemcpyToSymbol` function:

```
cudaMemcpyToSymbol(const char* var_name,
  const void *src, size_t count, size_t offset = 0,
  enum cudaMemcpyKind kind = cudaMemcpyHostToDevice)
```

This is essentially similar to the `cudaMemcpy` function presented in Chapter 3, but the first argument is the name of a variable whose storage has been specified with `__global__` or `__constant__`.

After the device has completed its operations, the host can access the content of a `__global__`/`__constant__` variable by calling `cudaMemcpyFromSymbol`:

```
cudaMemcpyFromSymbol(void* dst,
  const char* symbol, size_t count, size_t offset = 0,
  enum cudaMemcpyKind kind = cudaMemcpyDeviceToHost)
```

The following discussion demonstrates how these functions can be used.

Example Application – Adding Values in Different Memory Spaces

To demonstrate how these memory spaces can be accessed in code, the ch04_memory_spaces project adds values in shared and register memory and stores the results in global memory. The code in main.cu performs six operations:

1. Declare the array `sum` in global memory using the `__device__` specifier.
2. Declare the array `a` in constant memory using the `__constant__` specifier.
3. Initializes the values in the a array by calling `cudaMemcpyToSymbol`.
4. Launches the `addValues` kernel function, which adds the values of the `a` array (constant memory) to the value of `i` (register memory) and stores the result in `sum` (global memory).
5. Accesses the values of the `sum` array by calling `cudaMemcpyFromSymbol`.
6. Prints the values in the `sum` array to standard output.

Listing 4.2 presents the code in the ch04_memory_spaces/main.cu source file. A macro is used to set the array size and number of threads to 10.

Listing 4.2 ch04_memory_spaces/main.cu

```
#define NUM_VALS 10

// Store variable in global memory
__device__ int sum[NUM_VALS];

// Store variable in constant memory
__constant__ int a[NUM_VALS];

// Kernel function
__global__ void addVals() {
  int i = threadIdx.x;
  sum[i] = a[i] + i;
}

int main() {

  // Initialize content of a array
  int aData[NUM_VALS] = { 3, 3, 3, 3, 3, 3, 3, 3, 3, 3 };
  checkCudaErrors(cudaMemcpyToSymbol(a, aData,
    NUM_VALS * sizeof(int)));
```

Listing 4.2 ch04_memory_spaces/main.cu (Continued)

```
  // Execute kernel
  addVals<<<1, NUM_VALS>>>();

  // Access processed output
  int sumData[NUM_VALS];
  checkCudaErrors(cudaMemcpyFromSymbol(sumData, sum,
    NUM_VALS * sizeof(int)));

  // Display results
  std::cout << "Output: " << std::endl;
  for (int i = 0; i < NUM_VALS; i++) {
    std::cout << sumData[i] << std::endl;
  }
  return 0;
}
```

The host application doesn't pass any parameters to the kernel because the `sum` and `a` variables have been declared externally. Inside the kernel, the `i` variable doesn't have any specifiers, so the compiler stores it in register memory. Its value is set equal to a component of `threadIdx`, which is a built-in variable.

Thread Dimensions and Built-in Variables

So far, the example projects we've seen have sent one-dimensional data (arrays) to the device for processing. But in many applications, kernels need to access data in two or three dimensions. For example, if you want the GPU to filter an image, it's easier to work with a pixel's x and y coordinates instead of accessing a single dimension.

To make multi-dimensional data access possible, applications need to create thread blocks and grids with multiple dimensions. The dimensionality is set by invoking the kernel with special variables that identify the maximum size in each dimension.

For one-dimensional data access, the host application sets the size of thread blocks and grids using simple integers. The following code executes the kernel `foo` with three thread blocks in each grid and five threads in each thread block:

```
int grid_size = 3;
int block_size = 5;
foo<<<grid_size, block_size>>>(...);
```

To create grids and thread blocks that access multi-dimensional data, these values must be created with the `dim3` type, which accepts one, two, or three integer values. If one value is given, the `dim3` will represent one dimension. If two values are given, the dim3 will represent two dimensions, and if three values are given, the `dim3` will represent three dimensions.

An example will make this clear. Suppose the application wants to run foo with thread blocks in a 2-by-5 grid, where each thread block contains threads in a 3-by-4 array. Figure 4.1 illustrates how these grids and thread blocks are structured.

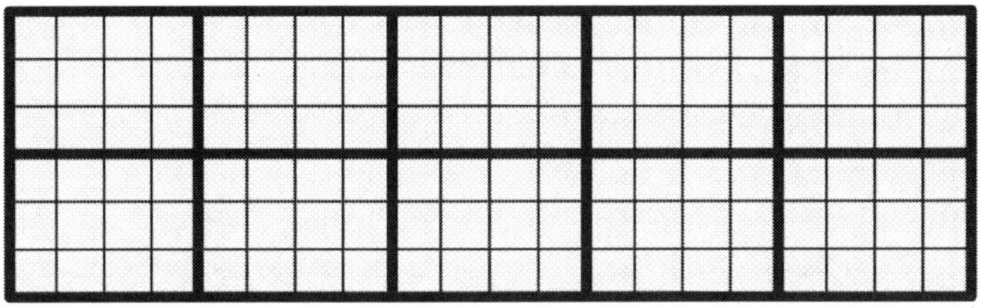

Figure 4.1 Two-Dimensional Grids and Thread Blocks

Each small square in this figure represents a thread, and because each thread block has dimensions 3-by-4, there are 12 threads per thread block. The grid's dimensionality is 2-by-5, so there are 10 thread blocks. This means the kernel will be executed by a total of $12 \times 10 = 120$ threads.

In code, this structure can be defined by creating two `dim3` values. The first identifies the grid's dimensionality and the second identifies the thread block dimensionality. These values can then be used when invoking the kernel.

```
// Set the dimensions of the grid
dim3 grid_dims(2, 5);

// Set the dimensions of each thread block in the grid
dim3 block_dims(3, 4);

// Invoke the kernel
foo<<<grid_dims, block_dims>>>(...);
```

Now each thread that executes `foo` will be able to access data using two dimensions. To be precise, each thread has a two-dimensional ID that it can use to access data. A thread can access its ID through the built-in `threadIdx` variable. Table 4.2 lists `threadIdx` and other variables that can be accessed in a kernel.

Table 4.2
Built-In Variables

Built-in Variable Name	Type	Description
`threadIdx`	`uint3`	Index of a thread inside a block
`blockIdx`	`uint3`	Index of a block inside a grid
`blockDim`	`dim3`	Dimensions of a thread block
`gridDim`	`dim3`	Dimensions of a grid
`warpSize`	`int`	Number of threads in a warp

An earlier discussion introduced vector types like `uint3` and `dim3` and explained how elements can be accessed with the subscripts `x`, `y`, and `z`. With this in mind, a one-dimensional thread can access its ID as `threadIdx.x` and a two-dimensional thread can access its ID with `threadIdx.x` and `threadIdx.y`.

It's important to note that `threadIdx` distinguishes a thread from other threads in the thread block, and doesn't uniquely identify the thread across the device. That is, threads in different thread blocks may have the same value for `threadIdx`.

The `blockIdx` variable is similar to `threadIdx`, but it provides the ID of a thread's *block*. Every thread in the same block will have the same `blockIdx` value. This becomes especially important when threads need to access shared memory because each thread block has its own shared memory region.

A thread can determine how many threads are being executed by accessing the `blockDim` and `gridDim` variables. If `blockDim` and `gridDim` are one-dimensional, then the total number of threads can be obtained by taking the product of the two values.

The last variable, `warpSize`, requires explanation. Threads in a thread block do not necessarily execute at the same time. But threads will execute simultaneously if they belong to the same *warp*. Warps are the basic units of execution on Nvidia devices, and their size is determined by the hardware, not the application.

In summary, a grid is a collection of thread blocks, a thread block is a collection of warps, and a warp is a collection of threads that execute at the same time. When setting the size of a thread block, it's good to ensure that it contains an integer number of warps.

Example Application – Printing Values of Built-In Variables

Nearly all the code in this book creates one-dimensional thread blocks. One exception is the ch04_print_builtin project, which creates a two-dimensional grid made up of two-dimensional thread blocks. It also demonstrates how built-in variables can be accessed.

The source code of the ch04_print_builtin project is in the main.cu source file. This code performs five operations:

1. Allocates a block of device memory to hold the application's output.
2. Defines the dimensions of the grid and the thread blocks in the grid.
3. Invokes the `printBuiltin` kernel function and passes the pointer to the allocated device memory.
4. Calls `cudaMemcpy` to read the data from the output memory block.
5. Displays the output data, which contains the values of the kernel's built-in variables.

Listing 4.3 presents the code in ch04_print_builtin/main.cu. The host application creates a two-dimensional grid of thread blocks and each thread block contains a two-dimensional array of threads.

Listing 4.3 ch04_print_builtin/main.cu

```
#define NUM_VALS 7

// The kernel function to be executed on the GPU
__global__ void printBuiltin(int* output) {
  if ((threadIdx.x == 0) && (threadIdx.y == 0) &&
     (blockIdx.x == 0) && (blockIdx.y == 0)) {
    output[0] = blockDim.x;
    output[1] = blockDim.y;
    output[2] = blockDim.z;
    output[3] = gridDim.x;
    output[4] = gridDim.y;
    output[5] = gridDim.z;
    output[6] = warpSize;
  }
}

int main() {

  // Allocate buffer for the output array
  int* output;
  checkCudaErrors(cudaMalloc(reinterpret_cast<void**>(&output),
    NUM_VALS * sizeof(int)));

  // Set the dimensions of the grid and thread blocks
  dim3 grid_dims(2, 5);
  dim3 block_dims(3, 4);
```

Listing 4.3 ch04_print_builtin/main.cu (Continued)

```
    // Invoke the kernel
    printBuiltin<<<grid_dims, block_dims>>>(output);

    // Access processed output
    int outputData[NUM_VALS];
    checkCudaErrors(cudaMemcpy(outputData, output,
      NUM_VALS * sizeof(int), cudaMemcpyDeviceToHost));

    // Display result
    std::cout << "blockDim.x: " << outputData[0] << std::endl;
    std::cout << "blockDim.y: " << outputData[1] << std::endl;
    std::cout << "blockDim.z: " << outputData[2] << std::endl;
    std::cout << "gridDim.x: " << outputData[3] << std::endl;
    std::cout << "gridDim.y: " << outputData[4] << std::endl;
    std::cout << "gridDim.z: " << outputData[5] << std::endl;
    std::cout << "warpSize: " << outputData[6] << std::endl;

    // Deallocate device memory
    cudaFree(output);

    return 0;
}
```

On my system, the application prints the following output:

```
blockDim.x: 3
blockDim.y: 4
blockDim.z: 1
gridDim.x: 2
gridDim.y: 5
gridDim.z: 1
warpSize: 32
```

These values should make sense. The grid's dimensionality is set to (2, 5), so `gridDim.x` equals 2 and `gridDim.y` equals 5. The dimensionality of each thread block is set to (3, 4), so `blockDim.x` equals 3 and `blockDim.y` equals 4.

The application didn't set values for the third dimension, but `blockDim.z` equals 1 and `gridDim.z` equals 1. This should make sense, since the default length of each dimension is always set to 1.

4.2 Processing Architecture

When you need to execute tasks on a GPU, it helps to understand how the device's resources and processing elements are structured. This allows you to plan where data should be stored and where routines should be processed to take full advantage of the device's architecture.

The fundamental processing element in an Nvidia GPU is the *streaming multiprocessor*, or *SM*. The number of SMs varies from device to device (the A100 has 108, the H100 has 144, and the latest Blackwell has 192), but they all have the same basic characteristics:

1. They send instructions to *warp schedulers*, which execute operations using threads. Each warp scheduler can access dedicated memory through its register file.
2. They process data using specialized cores called *CUDA cores*. The warp scheduler dispatches its instructions to CUDA cores, which can execute the same instruction on different data. This is SIMT (Single Instruction, Multiple Threads) processing.
3. Each SM has a dedicated block of on-chip memory that we refer to as shared memory. SMs can also access off-chip data in global memory and constant memory.

On Blackwell processors, each SM has four warp schedulers and each scheduler can access its own set of CUDA cores. Figure 4.2 provides a simplified view of how SMs in a Blackwell GPU are structured.

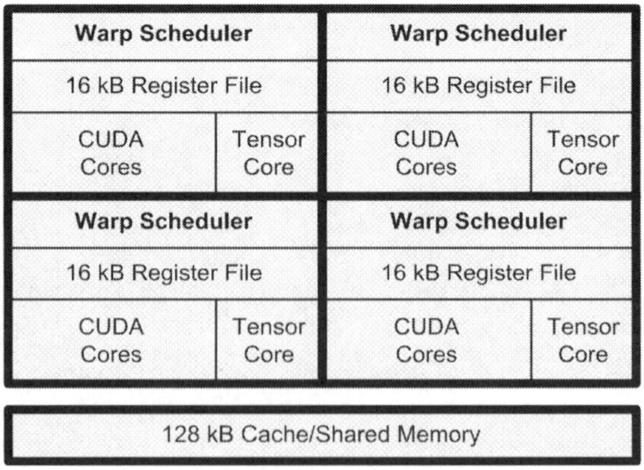

Figure 4.2 Simplified Structure of a Blackwell Streaming Multiprocessor (SM)

As illustrated, each warp scheduler in a Blackwell GPU can access its own tensor core, which accelerates matrix multiplication and accumulation operations. This is one of the many advanced features available on modern Nvidia devices, and others include ray tracing (RT) cores and texture units.

When a host application invokes a kernel, the driver assigns thread blocks to SMs. Depending on resource usage, an SM can execute multiple thread blocks, but a thread block is never split across multiple SMs. Within each thread block, threads are split into warps, which are directed to warp schedulers for processing.

It's important to see the difference between the register file and the shared memory. Each thread can access a portion of the register file that no other thread can access. However, every thread in the SM can access the same block of shared memory. But shared memory is SM-specific, so threads in different SMs can't access the same shared memory.

4.3 Device Model Analogy

To harness the maximum performance of an Nvidia GPU, you need to have a clear understanding of how code executes on the device. In particular, you need to store data where it can be accessed as quickly as possible. From fastest to slowest, the four types of memory are:

1. **Register** — storage specific for each thread
2. **Shared** — storage specific for each thread block
3. **Constant** — read-only storage for all threads
4. **Global** — read-write storage for all threads

To clarify how these memory types relate to threads and streaming multiprocessors, I've devised a simple analogy that compares threads in an SM to math students in a school. This section presents both parts of the analogy.

4.3.1 Analogy Part 1: Math Students in School

When I was in middle school, my math teacher would assign thirty problems to her thirty students every day, but she didn't check all 900 answers. Instead, she assigned a problem to each of us, and we'd go to the front of the class and copy our work from our notebook to the blackboard. The teacher would look over our solutions, and if she liked what she saw, we'd get a good grade.

An Nvidia device is like a school made up of these classrooms. Each room has students performing math problems. Each student has a notebook and students in a classroom share a blackboard. Students in the same class can work together at their blackboard, but students in different classes can't work together.

Every student in the school works on the same problem, but with different values. For example, if the problem involves adding two numbers, one student might add 1 + 2, another might add 3 + 4, and another might add 5 + 6. When all the students in a classroom finish, they can leave. Then the blackboard is erased and a new class of students will come in and work on the same problem, but with different values.

Each student entering the school knows what problem they'll be solving, but they don't know what values they'll be working with. The blackboard in each classroom is initially blank, so the students need to go to a central blackboard that contains values for the entire school. This central blackboard is much larger than the blackboards in the classrooms, but because of the long hallway, it takes quite a bit of time for students to read its values. Figure 4.3 illustrates the relationship between classes, classrooms, students, notebooks, and blackboards.

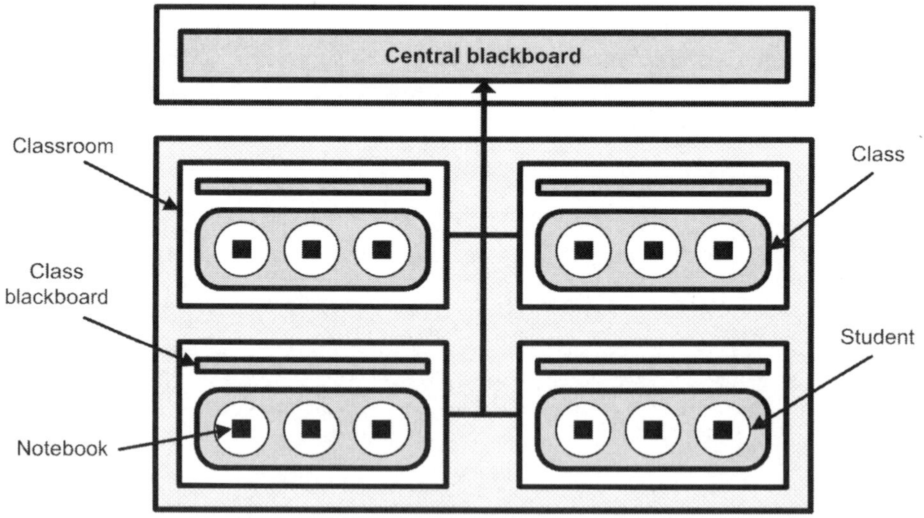

Figure 4.3 School of Math Students

In most cases, students will go to the central blackboard only twice—once to read the values for their problem and once to write down their answer. Because the central blackboard is so far away, students do their actual solving using their personal notebooks and classroom blackboards.

Students in different classes can't communicate, so the students in Class 1 won't know when the students in Class 2 have finished. The only way to be certain that a class has finished is when the schoolday ends. Once all of the answers have been written to the central blackboard, the schoolday is over.

It's important to make the distinction between classrooms and classes. A classroom is a physical area with a blackboard. A class is a group of students that occupy a classroom. When one class leaves a classroom, another can enter.

Each class has an ID that distinguishes it from every other class. Similarly, each student has an ID that distinguishes them from every other student in the class. Students don't have unique IDs across the entire school, but a global ID can be obtained by combining the Class ID and the Student ID.

4.3.2 Analogy Part 2: Threads in a Device

In my analogy, the school corresponds to a CUDA-compatible device and the math problem represents the kernel. Each classroom corresponds to a streaming multiprocessor (SM), and just as each classroom can be occupied by a class, each SM can be occupied by a thread block. Put simply, classes correspond to thread blocks and students correspond to threads. Figure 4.4 illustrates how threads and thread blocks are organized in an Nvidia device.

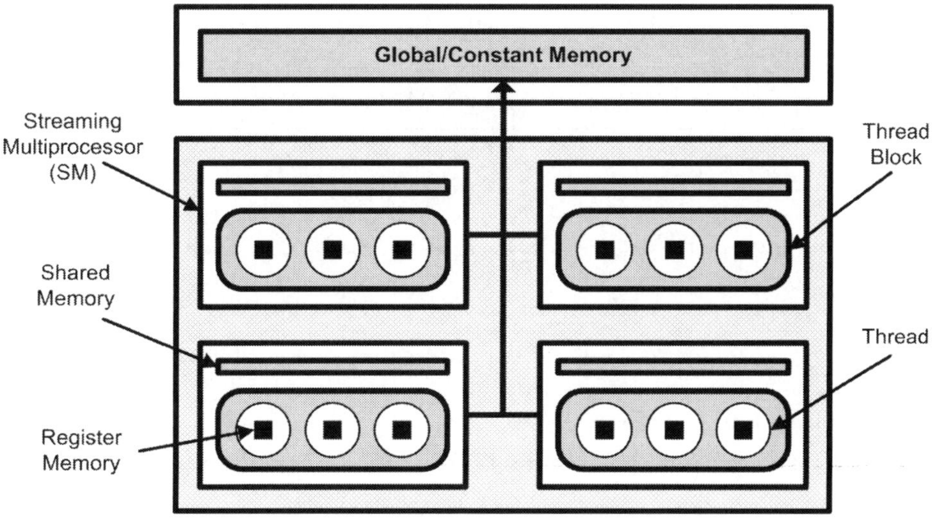

Figure 4.4 Threads in an Nvidia Device

In my analogy, the central blackboard corresponds to the device's global/constant memory, which can be accessed by the host and the device. When the host application copies data to the device, it's stored in global/constant memory. When the host copies data from a device, it comes from the device's global memory. Global/constant memory is the largest memory region, but it's also the slowest for threads to access.

Threads can access shared memory much faster than they can access global/constant memory, and this corresponds to the blackboards in each classroom. Shared memory isn't nearly as large as global/constant memory, but because of the speed, it's a good place for threads to store intermediate results during the execution of a kernel. Just as students in the same class can work together on the blackboard, threads in the same thread block can access the same block of shared memory. Threads in different thread blocks can never access the same shared memory.

The register memory corresponds to the notebook each student uses to solve a math problem. Each thread has exclusive access to its private memory, and it can access this faster than it can access any other type of memory. But this address space is much smaller than any other address space, so it's important not to use too much of it.

4.4 Shared Memory and Synchronization

When CUDA applications perform badly, the problem is usually caused by one of two mistakes:

- Using global memory for intermediate computation.
- Creating structures in register memory that are too large to fit.

In the second case, the compiler allocates memory from *local memory*, which is a portion of global memory used to store excess data. Global memory is very slow compared to other types of memory, so it's important to only access it when it's absolutely necessary.

In both cases, applications are usually better off storing data in shared memory, which is much faster than global memory. To use shared memory for intermediate computation, applications perform three steps:

1. Transfer input data from global memory to shared memory.
2. Process data in shared memory.
3. Transfer processed data from shared memory to the output block in global memory.

Looking at these steps, it's important to remember that host applications can't access shared memory in any way. Shared memory variables are created when a device starts executing a kernel and are freed when the device stops executing the kernel. This is why threads need to move input data from global memory to shared memory and then move processed data from shared memory to global memory.

This presents a problem. You don't want some threads to start processing data while other threads are still moving data from global memory to shared memory. For this reason, CUDA provides a function called `__syncthreads()`. When this is called in a kernel, every thread in the thread block will halt until every other thread in the thread block has reached the function call.

This function provides *block-level synchronization*, which means it only affects the threads in the same block as the current thread. It can synchronize operations involving shared memory but it can't synchronize operations involving global memory. The only way to be certain that threads have finished accessing global memory is to wait until the kernel has finished executing.

This presents a general structure for kernels that use shared memory during computation:

```
__global__ void foo(float* input_data, float* output_data) {
  __shared__ float shared_data[100];
  // Transfer data from input_data to shared_data
  __syncthreads();
  // Process data
  __syncthreads();
  // Transfer processed data to output_data
}
```

In this code, `shared_data` is hard-coded to contain 100 `float`s. Because the memory size is set in advance, this is *static shared memory*. But CUDA also supports *dynamic shared memory*, in which the size of the shared memory variable is determined programmatically.

Using dynamic shared memory is common because the desired amount may depend on the size of the input data, which may not be known in advance. In addition, highly efficient algorithms may want to allocate all the shared memory available, and this changes from device to device.

To access dynamic shared memory in code, two steps are needed. First, the variable declaration needs to be preceded by the `extern` keyword and the array brackets need to be empty. This is shown in the following code:

```
extern __shared__ float shared_data[];
```

Second, the kernel invocation needs to add a third parameter that identifies the size of shared memory to be allocated. For example, suppose the grid dimensions are set with `grid_dims`, the thread block dimensions are set with `block_dims`, and the shared memory size should be `shared_mem_size`. The application would invoke the kernel with code like the following:

```
foo<<<grid_dims, block_dims, shared_mem_size>>>(...);
```

The code in the ch04_shared_sync project demonstrates how dynamic shared memory can be allocated and accessed. The goal is to compute the sum of an array of values. The host application performs six operations:

1. Calls `cudaDeviceGetAttribute` to determine the maximum number of threads in a thread block. This is divided by two and called `MAX_THREADS`.
2. Calls `cudaMalloc` to allocate a block of device memory capable of storing `MAX_THREADS` values and a block of memory to hold the computed sum.
3. Calls `cudaMemcpy` to initialize the input values in device memory.
4. Invokes the kernel with a one-dimensional thread block that contains `MAX_THREADS` threads.
5. Calls `cudaMemcpy` to access the computed sum from the block in device memory.
6. Displays the computed sum and the actual sum.

When the kernel starts, each thread computes its global ID. Remember that `threadIdx` only provides an ID within the thread block. To compute the global ID, the thread needs to multiply the block ID by the block size and add its ID inside the block. The following code shows how this can be accomplished in an application that creates a one-dimensional grid composed of one-dimensional thread blocks:

```
int global_id = blockIdx.x * blockDim.x + threadIdx.x;
```

Listing 4.4 presents the code for the main.cu source file in the ch04_shared_sync project. This uses dynamic shared memory allocation, so the shared memory variable declaration starts with the `extern` keyword.

Listing 4.4 ch04_shared_sync/main.cu

```
__global__ void blockReduce(float* input, float* output) {
  extern __shared__ float shared_data[];

  int local_id = threadIdx.x;
  int global_id = blockIdx.x * blockDim.x + threadIdx.x;

  // Load input into shared memory
  shared_data[local_id] = input[global_id];

  // Wait until input has been loaded
  __syncthreads();

  // Perform reduction
  for (int i = blockDim.x/2; i > 0; i >>= 1) {
    if (local_id < i) {
      shared_data[local_id] += shared_data[local_id + i];
    }
    __syncthreads();
  }

  // Transfer result to global memory
  if (local_id == 0) {
    output[0] = shared_data[0];
  }
}

int main() {

  // Determine the max number of threads per block
  int dev, MAX_THREADS;
  cudaGetDevice(&dev);
  checkCudaErrors(cudaDeviceGetAttribute(&MAX_THREADS,
    cudaDevAttrMaxThreadsPerBlock, dev));
  MAX_THREADS /= 2;

  // Allocate memory for input/output memory blocks
  float *inputMem, *outputMem;
  checkCudaErrors(cudaMalloc(reinterpret_cast<void**>(&inputMem),
    MAX_THREADS * sizeof(float)));
  checkCudaErrors(cudaMalloc(reinterpret_
    cast<void**>(&outputMem), sizeof(float)));
```

Listing 4.4 ch04_shared_sync/main.cu (Continued)

```
  // Initialize input memory
  float* inputData = new float[MAX_THREADS];
  for (int i = 0; i < MAX_THREADS; i++) {
    inputData[i] = i * 1.0f;
  }
  checkCudaErrors(cudaMemcpy(inputMem, inputData,
    MAX_THREADS * sizeof(float), cudaMemcpyHostToDevice));

  // Invoke kernel
  blockReduce<<<1, MAX_THREADS,
    MAX_THREADS * sizeof(float)>>>(inputMem, outputMem);

  // Read computed sum
  float computedSum;
  checkCudaErrors(cudaMemcpy(&computedSum, outputMem,
    sizeof(float), cudaMemcpyDeviceToHost));

  // Display and check results
  std::cout << "Computed sum: " << computedSum << std::endl;
  std::cout << "Actual sum: "
    << ((MAX_THREADS / 2) * (MAX_THREADS - 1)) << std::endl;

  cudaFree(inputMem);
  cudaFree(outputMem);
  delete[] inputData;
  return 0;
}
```

When writing professional code, it's good to check how much shared memory is available per block. This can be accomplished by calling `cudaDeviceGetAttribute` with the second parameter set to `cudaDevAttrMaxSharedMemoryPerBlock`. In the example code, the application assumes that the amount of memory is sufficient to hold the input data. But in the real world, this shouldn't be assumed.

Most of this code should look familiar, but the kernel invocation statement deserves explanation. The code is given as:

```
blockReduce<<<1, MAX_THREADS, MAX_THREADS * sizeof(float)>>>
  (inputMem, outputMem);
```

In this code, the last argument in the angle brackets sets the size of shared memory. The shared memory needs to hold every value in the input array, so its size is set to `MAX_THREADS` times the number of bytes in each `float`.

4.5 Summary

If CUDA's libraries don't provide a function that meets your need, you'll have to write your own kernel. This requires understanding of the device's memory spaces (global, shared, constant, register) and processing elements (threads, thread blocks, and grids). You also have to be familiar with new data types, built-in variables, and synchronization.

The best way to understand CUDA kernels is to compare them with C functions. Like a C function, a kernel function accepts parameters, declares variables, and executes statements in curly braces. Unlike C functions, kernels have no return value, their declarations start with `__global__`, and they can access different types of device memory. Of course, the central difference is that kernels execute across many threads and thread blocks.

When coding kernels, it's important to have a solid grasp of threads, thread blocks, and the memory spaces available. To help clarify matters, this chapter has presented an analogy that compares threads to students, thread blocks to classrooms, and shared memory to the blackboard inside each classroom.

Shared memory is much faster than global memory, and should therefore be used instead of global memory whenever possible. Each thread block has its own shared memory that its threads can access. But when accessing this memory, it's important to use proper synchronization to ensure that processing only continues after each thread is done using shared memory.

Chapter 5

Advanced Development Topics

Chapters 3 and 4 covered the fundamentals of CUDA programming, and if you grasped the content of these chapters, you'll have no trouble accessing the CUDA libraries discussed in later chapters. This chapter presents five topics that are useful and interesting but not absolutely necessary:

1. **Streams** — improve performance by making it possible to execute commands asynchronously
2. **Events** — used for synchronization and measurement of elapsed time
3. **Pinned and mapped memory** — improves performance of host-device data transfers
4. **Dynamic Parallelism** — Enables parent kernels to execute child kernels
5. **Tensor Cores and the WMMA API** — Uses advanced processing elements to perform Warp Matrix Multiply and Accumulate (WMMA) operations

This chapter discusses each of these topics and presents example code that demonstrates how they can be implemented. Streams and events are particularly useful, and are employed by some library functions to manage execution.

With each new generation of devices, Nvidia is becoming more and more focused on accelerating machine learning. This is reflected in the WMMA API, which makes it possible to perform matrix multiply and accumulate (MMA) using an entire warp at once. In addition to this, Appendix B presents the WGMMA (Warp Group Matrix Multiply and Accumulate) API, which performs MMA operations using a group of warps.

5.1 Streams

The word *stream* is used frequently in software development, and its meaning can change dramatically depending on the context. When it comes to CUDA development, a stream is a queue of device commands. A *command* is a request for the device to process data or transfer data from one location to another.

Every application can access a default stream, and by default, commands sent to the device are all contained in this stream. Because they belong to the same stream, the commands execute in strict order. That is, the first command executes first, the second command starts when the first is finished, and so on.

Using the default stream makes execution predictable, but modern GPUs are capable of executing multiple commands at once. To take advantage of this, applications can create streams in addition to the default stream. Commands in different streams can execute at the same time.

In code, the process of using a stream consists of five steps:

1. Declare a `cudaStream_t` structure.
2. Call `cudaStreamCreate` to initialize the `cudaStream_t`.
3. Use the `cudaStream_t` in kernel execution statements, data transfer functions, or any other function that accepts a stream.
4. Call `cudaStreamSynchronize` to wait for the stream to complete its processing.
5. When the stream is no longer needed, call `cudaStreamDestroy` to destroy its resources.

If commands belong to different streams, there's no way to know which will execute first. Streams can be synchronized using *events*, which will be discussed shortly.

As explained in Chapter 4, a kernel execution statement identifies the size of the grid, the size of each thread block, and the amount of shared memory to allocate. If the grid size is `grid_dims`, the thread block size is `block_dims`, and the desired shared memory is `mem_size`, the following statement will execute `foo`:

```
foo<<<grid_dims, block_dims, shared_mem_size>>>(...);
```

To execute `foo` in a stream named `stream1`, the stream's structure needs to be inserted inside the angle brackets. The following statement shows what this looks like:

```
foo<<<grid_dims, block_dims, shared_mem_size, stream1>>>(...);
```

This tells the device to execute `foo` in the stream represented by `stream1`. As a result, the kernel's execution may overlap with commands from other streams, such as the commands in the default stream.

Many CUDA functions accept a `cudaStream_t` and add a device command to the stream. Because the commands execute asynchronously with the default stream, the functions' names end with `Async`. Table 5.1 lists five functions that tell the device to transfer data asynchronously.

Table 5.1

Asynchronous Data Transfer Functions

Function	Description

`cudaMemcpyAsync(void* dst,`
`const void* src, size_t count,`
`cudaMemcpyKind kind,`
`cudaStream_t stream)`
	Copies data asynchronously from src to dst

`cudaMemsetAsync(void* mem, int value,` `size_t count, cudaStream_t stream)`	Asynchronously sets bytes in mem to the given value

`cudaMemcpyToSymbolAsync(`
`const void* symbol, const void* src,`
`size_t count, size_t offset,`
`cudaMemcpyKind kind,`
`cudaStream_t stream)`
	Copies asynchronously into a device symbol (variable preceded by __device__ or __constant__)

`cudaMemcpyFromSymbolAsync(void* dst,`
`const void* symbol, size_t count,`
`size_t offset, cudaMemcpyKind kind,`
`cudaStream_t stream)`
	Copies asynchronously from a device symbol (variable preceded by __device__ or __constant__)

`cudaMemcpyBatchAsync(`
`const void** dsts, const void** srcs,`
`const size_t* sizes, size_t count,`
`cudaMemcpyAttributes* attrs,`
`size_t* attrsIdxs, size_t numAttrs,`
`cudaStream_t stream)`
	Executes a batch of copy commands asynchronously

There's no point using streams if an application only needs to perform one task at a time. They become helpful when a device needs to perform several tasks, and some of the tasks can be executed in parallel.

The more tasks that can be executed asynchronously, the better performance these functions provide. But there are many instances where commands in different streams need to be synchronized. This is accomplished using *events*.

5.2 Events

In CUDA, an event is a marker in a stream that can be accessed in code. Applications can use these markers in three main ways:

1. Applications can query an event to check if the device has completed the commands preceding the event.
2. Applications can halt processing until the commands preceding an event have been completed.
3. If two events belong to the same stream, they can be used to determine how much time elapsed between the start of the first command following the first event and the completion of the last command preceding the second event.

Events can be hard to grasp, so it helps to look at an illustration. Figure 5.1 depicts a stream containing five commands, starting with the execution of kernel `foo`. The application inserted two events into the stream named Event 0 and Event 1.

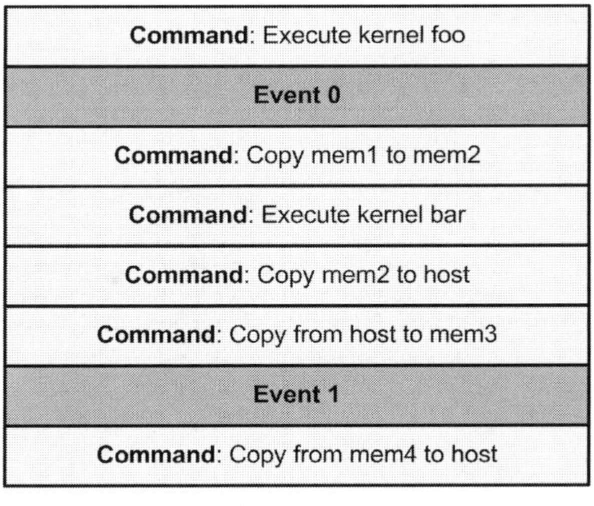

Figure 5.1 Two Events in a Stream

With these events in place, an application can check to see if the commands preceding Event 0 or Event 1 have executed. An application can also halt the CPU until the commands preceding an event have completed. It's also possible to determine the amount of time the GPU used to process the commands between Event 0 and Event 1. The goal of this section is to explain how these operations can be performed in code.

5.2.1 The Event Structure

Just as streams are represented by `cudaStream_t` structures, events are represented by `cudaEvent_t` structures. The general process of working with event structures consists of four steps:

1. Call `cudaEventCreate` to initialize the event structure.
2. Call `cudaEventRecord` with the `cudaEvent_t` and a `cudaStream_t` to insert the event into a stream.
3. Use the event in one or more operations involving queries, synchronization, or time profiling.
4. When the event is no longer needed, call `cudaEventDestroy` to free the structure and its resources.

The following discussions explain how events can be used. Afterward, I'll present an example application that uses events to synchronize streams.

5.2.2 Queries

After an event has been inserted into a stream, applications can easily check if the commands preceding it have completed. The function to call is `cudaEventQuery`, and its signature is given as follows:

```
cudaEventQuery(cudaEvent_t event)
```

This returns immediately with a value that identifies the status of the operations preceding the event. If the operations completed successfully, the function will return `cudaSuccess`, represented by `0`. If the operations haven't completed, the function will return `cudaErrorNotReady`, represented by `600`.

5.2.3 Synchronization

Events make two forms of synchronization possible. First, applications can force the CPU to halt processing until the device has completed executing the commands preceding the event. This is accomplished by calling `cudaEventSynchronize`, whose signature is given as follows:

```
cudaEventSynchronize(cudaEvent_t event)
```

In addition, an event in one stream can be used to halt processing in another stream. This is made possible by `cudaStreamWaitEvent`:

```
cudaStreamWaitEvent(cudaStream_t stream, cudaEvent_t event,
  unsigned int flags = 0)
```

When an application calls this, future commands in the stream given by `stream` won't be executed until the commands preceding `event` have executed in `event`'s stream. The `flags` argument can be set to one of two values:

- `cudaEventWaitDefault` — only events in the device context are supported
- `cudaEventWaitExternal` — events external to the device context can be used for synchronization

The project presented at the end of this section demonstrates how an application can use `cudaStreamWaitEvent` to synchronize operations in different streams.

5.2.4 Time Profiling

If two events have been inserted into the same stream, an application can access the elapsed time between them by calling the `cudaEventElapsedTime` function:

```
cudaEventElapsedTime(float* time, cudaEvent_t start,
  cudaEvent_t stop)
```

When an application calls this, the elapsed time between the `start` event and the `stop` event is stored in the `time` argument. Keep in mind that the time is given in milliseconds.

It's a good idea to synchronize the application with `cudaEventSynchronize` so that it waits for the stop event. The following code gives an idea of how this works:

```
cudaEvent_t start, stop;
cudaEventCreate(&start);
cudaEventCreate(&stop);
...
cudaEventRecord(start);
...
cudaEventRecord(stop);
...
cudaEventSynchronize(stop);
cudaEventElapsedTime(&time, start, stop);
```

5.2.5 Example Application

The ch05_stream_events project demonstrates how streams and events can work together. In particular, it shows how an event in one stream can be used to synchronize the processing of another stream. The project's code performs eight operations:

1. Creates two streams named `stream1` and `stream2`.
2. Creates an event named `event`.
3. Adds a command to `stream1` to execute the kernel named `kernel1`.
4. Calls `cudaEventRecord` to insert `event` into `stream1`.
5. Calls `cudaStreamWaitEvent` to halt `stream2`'s command execution until the commands preceding `event` are finished.
6. Adds a command to `stream2` to execute the kernel named `kernel2`.
7. Calls `cudaStreamSynchronize` to force the CPU to halt until both streams have finished executing their commands.
8. Frees the resources allocated by the application.

As a result of the synchronization, `stream2` won't execute `kernel2` until the commands before event in `stream1` have completed. This means `kernel1` will always execute before `kernel2` even though their commands belong to different streams.

The code that performs these operations is in the ch05_stream_events/main.cu source file. On the following page, Listing 5.1 presents the file's content.

Unlike preceding applications, the kernels don't process any data. Instead, each kernel calls `printf` to display a message identifying itself. This function is very convenient for kernel debugging, and during execution, the device performs three main operations:

- The device stores characters in a special buffer reserved for `printf` output.
- When the kernel ends or the host synchronizes, data from the buffer is transferred from the device to the host.
- The CUDA runtime receives the buffer and writes its content to the host's standard output stream.

Because of all the processing, `printf` should only be used for kernel debugging. It shouldn't be included in production code.

By default the `printf` buffer on the device is fairly small (about 1 MB). An application can increase its size by calling `cudaDeviceSetLimit` with the name of the limit set to `cudaLimitPrintfFifoSize`.

Listing 5.1 ch05_stream_events/main.cu

```
int main() {

  // Create two streams
  cudaStream_t stream1, stream2;
  checkCudaErrors(cudaStreamCreate(&stream1));
  checkCudaErrors(cudaStreamCreate(&stream2));

  // Create event
  cudaEvent_t event;
  checkCudaErrors(cudaEventCreate(&event));

  // Launch Kernel 1 in Stream 1
  kernel1<<<1, 1, 0, stream1>>>();

  // Record event in Stream 1 after Kernel 1
  checkCudaErrors(cudaEventRecord(event, stream1));

  // Make Stream 2 wait until event completes
  checkCudaErrors(cudaStreamWaitEvent(stream2, event, 0));

  // Launch Stream 2 in Stream 2
  kernel2<<<1, 1, 0, stream2>>>();

  // Wait until streams complete commands
  checkCudaErrors(cudaStreamSynchronize(stream1));
  checkCudaErrors(cudaStreamSynchronize(stream2));

  // Free event
  checkCudaErrors(cudaEventDestroy(event));

  // Free streams
  checkCudaErrors(cudaStreamDestroy(stream1));
  checkCudaErrors(cudaStreamDestroy(stream2));

  return 0;
}
```

Neither kernel uses shared memory, but the kernel execution statements need to identify the amount of shared memory to allocate. This explains why the third argument in each statement is `0`.

In both cases, the kernels are executed with one thread block, and the thread block contains a single thread. Keep in mind that stream synchronization affects every thread of every thread block in the stream.

5.3 Pinned and Mapped Memory

Operating systems allow programs to access more memory than is available in RAM. To make this possible, they divide the program's memory into blocks called *pages*, and transfer pages from storage to RAM as they become needed. Paging is convenient but reduces performance due to the need for repeated data transfers.

Many operating systems support *pinned* memory (also called *page-locked* memory), which is a region of memory that can't be transferred to and from secondary storage. Data stored in pinned memory can be transferred to a device faster than data in regular memory because it will never be swapped to and from disk.

In code, applications can allocate pinned memory by calling `cudaHostAlloc`, whose signature is given as follows:

```
cudaHostAlloc(void** mem, size_t size, unsigned int flags)
```

If this completes successfully, the first argument will point to a block of pinned memory whose size is given by the `size` parameter. The `flags` parameter can be set to one of the following values or an OR'ed combination:

- `cudaHostAllocDefault` — the pinned memory can only be accessed by the current device context (default)
- `cudaHostAllocPortable` — the pinned memory can be accessed by all devices
- `cudaHostAllocMapped` — enables the pinned memory to be associated with a block of memory on the device (mapped)
- `cudaHostAllocWriteCombined` — allocates write-combined memory

If an application uses default allocation (set `flags` to 0), the pinned memory can be used in `cudaMemcpy` like a regular host pointer. The following code demonstrates this:

```
// Allocate host memory
float* hostMem;
size_t size = 1024 * sizeof(float);
cudaHostAlloc((void**)&hostMem, size, 0);

// Allocate device memory
float* deviceMem;
cudaMalloc((void**)&deviceMem, size);

// Copy from host to device
cudaMemcpy(deviceMem, hostMem, size, cudaMemcpyHostToDevice);
```

If a device supports *mapped memory*, then a region of pinned memory on the host can be associated with a block of memory on the device. Creating this association is called *mapping*, and if pinned memory has been mapped to device memory, the GPU can transfer data using direct memory access (DMA). This means applications can access device memory without needing to call `cudaMemcpy`.

In code, the process of mapping a region of pinned memory to device memory requires three steps:

1. Enable mapped memory by calling `cudaSetDeviceFlags` with the argument set to `cudaDeviceMapHost`.
2. Allocate a block of mappable pinned memory by calling `cudaHostAlloc` with the `flags` parameter set to `cudaHostAllocMapped`.
3. Call `cudaHostGetDevicePointer` to obtain a pointer to device memory mapped to the pinned memory.

The last step requires explanation. Applications normally obtain pointers to device memory by calling `cudaMalloc`, but if mapping has been enabled and the application has allocated pinned memory that can be mapped, it can call `cudaHostGetDevicePointer` to obtain a pointer to mapped device memory. The function's signature is:

```
cudaHostGetDevicePointer(void** deviceMem, void* hostMem,
  unsigned int flags)
```

After this executes, the `deviceMem` pointer can be used to access device memory as though `cudaMalloc` had been called. There's no need to set the size of the allocation because it must have the same size as the pinned memory referenced by `hostMem`. For now, the `flags` parameter must be set to 0.

There are three main advantages to obtaining pointers to mapped device memory with `cudaHostGetDevicePointer` instead of obtaining regular device pointers with `cudaMalloc`:

1. Because of the mapping from host memory to device memory, data transfers are faster because of direct memory access.
2. In code, applications don't have to use `cudaMemcpy` to transfer data to and from mapped device memory. They can access the mapped memory on the host instead.
3. Because of the mapping, the device memory pointer doesn't have to be deallocated with `cudaFree`. The block of memory is deallocated when the host memory is deallocated with the function `cudaFreeHost`.

The ch05_mapped_memory project provides a simple demonstration of how mapped memory can be used. It stores an array of values in pinned memory and then obtains a device pointer to mapped memory on the device. To be precise, the code in ch05_mapped_memory/main.cu performs six operations:

1. It enables memory mapping by calling `cudaSetDeviceFlags` with `cudaDeviceMapHost`.
2. It calls `cudaHostAlloc` to allocate pinned memory on the host. Because the `cudaHostAllocMapped` flag is set, this memory can be mapped to device memory.
3. It calls `cudaHostGetDevicePointer` to obtain a pointer to the mapped memory on the device.
4. It launches the kernel, which scales values of an array.
5. It calls `cudaDeviceSynchronize` to ensure that the device has finished executing the kernel.
6. It accesses the processed data through the pinned memory on the host and then frees the pinned memory.

Listing 5.2 presents the code in ch05_mapped_memory/main.cu. As shown, there are no calls to `cudaMemcpy` or `cudaFree`.

Listing 5.2 ch05_mapped_memory/main.cu

```
int main() {
  const int numVals = 16;
  const size_t size = numVals * sizeof(int);

  // Enable mapped memory
  cudaSetDeviceFlags(cudaDeviceMapHost);

  // Allocate mappable pinned memory array on host
  float *hostMem;
  checkCudaErrors(cudaHostAlloc(
    reinterpret_cast<void**>(&hostMem), size,
    cudaHostAllocMapped));

  // Initialize host memory
  for (int i=0; i<numVals; i++) {
    hostMem[i] = 1.0f * i;
  }
```

Listing 5.2 ch05_mapped_memory/main.cu (Continued)

```
  // Obtain pointer to device memory mapped to host memory
  float* deviceMem;
  checkCudaErrors(cudaHostGetDevicePointer(
    reinterpret_cast<void**>(&deviceMem),
    reinterpret_cast<void*>(hostMem), 0));

  // Launch kernel
  const float scalingFactor = 7.0f;
  scaleArray<<<1, numVals>>>(deviceMem, scalingFactor);

  // Synchronize to make sure kernel finished
  checkCudaErrors(cudaDeviceSynchronize());

  // Display processed data
  std::cout << "Results:" << std::endl;
  for (int i=0; i<numVals; i++) {
    std::cout << hostMem[i] << " ";
  }
  std::cout << std::endl;

  // Free host memory
  checkCudaErrors(cudaFreeHost(hostMem));

  return 0;
}
```

In this code, the kernel executes with 16 threads and each thread multiplies an element of the input array by the scaling factor (7.0). Before printing the results, the application calls `cudaDeviceSynchronize`, which forces the CPU to wait until all of the threads have finished.

After the kernel is finished, there's no need to call `cudaMemcpy` to transfer the processed data to the host. Because of the memory mapping, the application can access the processed data through the pointer to host memory (`hostMem`).

Unlike preceding examples, the code in this application doesn't call `cudaFree` to deallocate the pointer to device memory. Instead, it calls `cudaFreeHost` to free the pinned memory region on the host. When this is called, the mapped region of device memory will automatically be deallocated.

It's important to clearly understand the difference between pinned memory and mapped memory. Pinned memory is host memory that won't be swapped to and from disk storage. Mapped memory is pinned memory that has been associated with a region of device memory.

5.4 Dynamic Parallelism

In most applications, kernel execution is controlled by the host program. The program decides which kernels should be executed and when they should be executed. This is ideal when the amount of work to be performed is known in advance.

But in some cases, applications may not know in advance how much work needs to be performed by the GPU. For this reason, CUDA makes it possible for one kernel to execute another, and this capability is called *dynamic parallelism*.

The main advantage of dynamic parallelism is that it reduces the time spent synchronizing between the CPU and GPU. When one kernel on the device launches another kernel on the device, the synchronization can be performed on the device without alerting the CPU.

The main disadvantage is that dynamic parallelism can take a significant amount of time and processing resources. This is due to the processing overhead of launching a kernel and the time needed to transfer data. For this reason, it's a good idea to test your code to see if dynamic parallelism will provide acceptable performance.

At minimum, the process of using dynamic parallelism involves one kernel (the *parent*) invoking a second kernel (the *child*). This uses essentially the same code that regular applications use, but the code is moved to the kernel. There are three main steps involved:

1. In the host application, provide code for the parent and child kernels.
2. In the parent kernel, include a statement that invokes the child kernel.
3. When building the application, set the relocatable device code option (`rdc`) to true. On the command line, this can be done with `-rdc=true`.

Algorithms involving recursion are well-suited for dynamic parallelism because kernels need to call themselves. This is demonstrated in the ch05_tree_recursion project, which traverses a tree of nodes, where each node has an integer value, an integer number of children, and an array of integers that represent child nodes. In code, each node is represented by the `Node` structure, which is defined in the following way:

```
struct Node {
  int value;
  int numChildren;
  int* children;
};
```

To demonstrate tree traversal with dynamic parallelism, the ch05_tree_recursion project creates a tree containing five nodes. Figure 5.2 shows what this looks like.

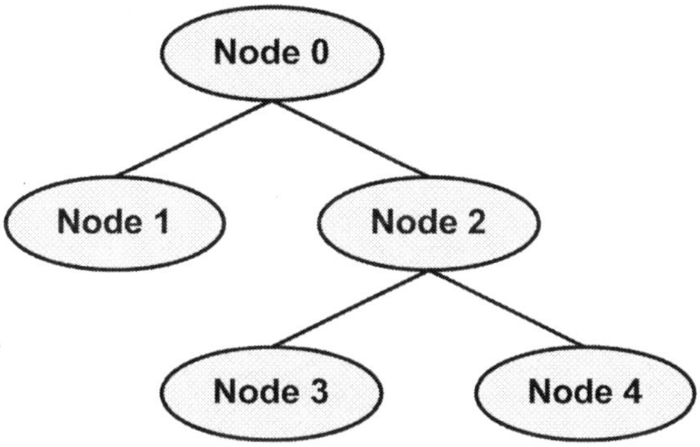

Figure 5.2 Example Tree for Dynamic Parallelism

Node 0 has two children and Node 2 has two children, so `numChildren` equals 2 for these nodes and 0 for the others. The children of Node 0 are Node 1 and Node 2, so its `children` field will be set to `{0, 1}`. Similarly, the `children` field of Node 2 will be set to `{3, 4}`.

The source code of the ch05_tree_recursion project can be found in main.cu. If you look through this file, you'll see that it performs seven tasks:

1. Create a block of device memory for each node's child array.
2. Copy the child indexes into each block of device memory.
3. Create and initialize the array of `Node` structures on the host.
4. Create a block of device memory for the `Node` structures.
5. Copy the host `Node` structures to the block of device memory.
6. Launch the `processNode` kernel with one thread and one thread block.
7. Wait for the kernel to finish, and then free the blocks of device memory.

The `processNode` kernel starts by printing its index, depth, and value. Then it iterates through its children and calls `processNode` once for each child. The depth starts at 0 and increases by 1 each time a child node is processed.

Listing 5.3 presents the code in ch05_tree_recursion/main.cu. When building this code, keep in mind that the compiler's relocatable device code option (`rdc`) must be set to true. If this isn't set, you'll get an error about the kernel launch requiring a separate compilation mode.

Listing 5.3 ch05_tree_recursion/main.cu

```
// Kernel to process nodes
__global__ void processNode(Node* nodes, int index, int depth) {
  Node node = nodes[index];

  // Print node index, depth, and value
  printf("Processing node %d at depth %d with value %d\n",
    index, depth, node.value);

  // Launch kernel for each child
  for (int i = 0; i < node.numChildren; i++) {
    int childIndex = node.children[i];
    processNode<<<1, 1>>>(nodes, childIndex, depth + 1);
  }
}

int main() {

  // The number of nodes in the tree
  const int numNodes = 5;

  // Create arrays identifying children
  int children0[2] = { 1, 2 };
  int children2[2] = { 3, 4 };
  int *deviceChildren0, *deviceChildren2;
  checkCudaErrors(cudaMalloc(&deviceChildren0, 2 * sizeof(int)));
  checkCudaErrors(cudaMalloc(&deviceChildren2, 2 * sizeof(int)));

  // Copy node children arrays to device
  checkCudaErrors(cudaMemcpy(deviceChildren0,
    children0, 2 * sizeof(int), cudaMemcpyHostToDevice));
  checkCudaErrors(cudaMemcpy(deviceChildren2, children2,
    2 * sizeof(int), cudaMemcpyHostToDevice));

  // Node data
  Node hostNodes[numNodes];
  hostNodes[0] = { 0, 2, deviceChildren0 };
  hostNodes[1] = { 1, 0, nullptr };
  hostNodes[2] = { 2, 2, deviceChildren2 };
  hostNodes[3] = { 3, 0, nullptr };
  hostNodes[4] = { 4, 0, nullptr };
```

Listing 5.3 ch05_tree_recursion/main.cu (Continued)

```
    // Copy node data to device
    Node* deviceNodes;
    checkCudaErrors(cudaMalloc(&deviceNodes,
      numNodes * sizeof(Node)));
    checkCudaErrors(cudaMemcpy(deviceNodes, hostNodes,
      numNodes * sizeof(Node), cudaMemcpyHostToDevice));

    // Launch kernel
    processNode<<<1, 1>>>(deviceNodes, 0, 0);

    // Wait for kernel to execute
    cudaDeviceSynchronize();

    // Free device memory
    cudaFree(deviceNodes);
    cudaFree(deviceChildren0);
    cudaFree(deviceChildren2);

    return 0;
}
```

In general, algorithms that require recursion aren't commonly run on GPUs due to the unpredictable control paths. But if your program requires one kernel to call another, dynamic parallelism will make this possible.

5.5 Tensor Cores and the WMMA API

One of the most important improvements in Nvidia's Volta architecture was the introduction of *tensor cores*, which make it possible to perform matrix multiply-and-accumulate (MMA) operations at high speed. These operations play a central role in machine learning, which is why Nvidia markets the Volta architecture for model training and inference.

If high-speed machine learning is your goal, you're better off using a high-level toolset like PyTorch or TensorFlow. But if you have a custom algorithm that requires MMA, it's a good idea to learn about the WMMA (Warp Matrix Multiply Accumulate) API. This provides data structures and functions that make it possible to access tensor cores directly.

Before I delve into the details of the WMMA API, I'd like to briefly explain the difference between thread blocks and warps:

1. Threads in a thread block can access the same shared memory region and can be synchronized together. Applications can set the size of a thread block up to the maximum supported size. Threads in a thread block do not necessarily execute at the same time.
2. Threads in a warp execute the same instructions in parallel. The size of a warp is fixed by the device, and for many devices, every warp has 32 threads.

One major difference between the WMMA API and regular CUDA is that WMMA functions involve all the threads in an entire warp at once. For this reason, applications need to set the thread block size to a multiple of the warp size (usually 32).

This section introduces the data structures and functions in the WMMA API. It ends by walking through example code that multiplies two 128-by-128 matrices together and adds a third matrix.

5.5.1 Fragments

In the CUDA Toolkit, the mma.h header declares the `wmma` namespace, which contains the elements of the WMMA API. The API's central data structure is the `fragment`, which stores a block of data to be used in matrix operations.

To understand fragments, it's important to be familiar with some terminology. A *tile* is a block of data that can be processed by a single warp. A *matrix* is a mathematical structure made up of one or more tiles (Chapter 6 discusses matrices and matrix operations in detail).

The goal of WMMA is to compute Matrix D = Matrix A × Matrix B + Matrix C. Matrix A and Matrix B don't have to be the same size, but the number of columns in Matrix A has to equal the number of rows in Matrix B. If Matrix A has dimensions M-by-K and Matrix B has dimensions K-by-N, Matrix C and Matrix D will have dimensions M-by-N. In this example, K is called the *inner dimension*. WMMA refers to Matrix C as the *accumulator*.

Inside a kernel, each matrix is represented by a fragment. To declare a fragment, the kernel needs to provide six template arguments:

1. Which matrix in the MMA operation will contain the fragment's data (`matrix_a`, `matrix_b`, or `accumulator`).
2. The number of rows in each tile.
3. The number of columns in each tile.

4. The size of the inner dimension.
5. The data type of the elements stored in the matrices (`half`, `float`, `signed char`, `unsigned char`, and others).
6. For fragments in Matrix A or Matrix B, the layout of the values in the fragment must be set to `wmma::row_major` or `wmma::col_major`.

It's important to see that the dimensions given in the declaration are the dimensions of the *tiles* that make up a matrix, and not the dimensions of the matrix. In general, it's safe to set the three dimensions to (16, 16, 16). For optimal performance, these dimensions should be configured to reflect the target hardware.

The last template argument identifies how the values in Matrix A and Matrix B are arranged. If this is set to `wmma::row_major`, the elements of each row of the matrix are stored contiguously. That is, the first element of Row N + 1 follows the last element of Row N. If this is set to `wmma::column_major`, the elements of each column are stored contiguously. This means the first element of Column N + 1 follows the last element of Column N.

To better understand fragments, it helps to see how they can be declared in a kernel. If the tile dimensions are `WMMA_M`, `WMMA_N`, and `WMMA_K`, the following code declares fragments for Matrix A, Matrix B, and Matrix C of the MMA operation:

```
// Declare a fragment in Matrix A
wmma::fragment<wmma::matrix_a, WMMA_M, WMMA_N, WMMA_K,
  half, wmma::row_major> a_frag;

// Declare a fragment in Matrix B
wmma::fragment<wmma::matrix_b, WMMA_M, WMMA_N, WMMA_K,
  half, wmma::col_major> b_frag;

// Declare a fragment in Matrix C
wmma::fragment<wmma::accumulator, WMMA_M, WMMA_N, WMMA_K,
  float> c_frag;
```

It's important to notice the types used for the different matrices. Matrix A and Matrix B contain half-precision values (`half`), and Matrix C contains single-precision values (`float`). This is called *mixed-precision computation*, and it's important to understand.

Half-precision values make some developers nervous, but tensor cores are optimized to process matrices containing half-precision values at high speed. But the final MMA operation relies on single-precision values, which increases accuracy and reduces the likelihood of saturation and the loss of significant digits.

5.5.2 Fragment Functions

In addition to providing the `wmma::fragment` structure, the mma.h header provides a handful of functions that access and operate on fragments. Table 5.2 lists these functions and provides a description of each.

Table 5.2

Fragment Functions

Function	Description
`fill_fragment(fragment& a, const T& v)`	Fills a fragment with a single value
`load_matrix_sync(fragment& a, const T* devptr, unsigned stride)`	Loads data from a device pointer
`load_matrix_sync(fragment& a, const T* devptr, unsigned stride, layout_t layout)`	Loads data from a device pointer with the given layout
`store_matrix_sync(T* devptr, const fragment& a, unsigned stride, layout_t layout)`	Transfers data from a fragment to a device pointer with the given layout
`mma_sync(fragment& d, const fragment& a, const fragment& b, const fragment& c, bool satf = false)`	Performs the matrix multiply and accumulate (MMA) operation

The first function, `fill_fragment`, is the easiest to understand. It sets every element of the fragment to the value set in the second argument. The value must have the same type as the fragment, so if the fragment was declared to contain half-precision values, the second argument of `fill_fragment` must be a half-precision value.

Most applications initialize fragments using data stored in blocks of device memory. This is accomplished with `load_matrix_sync`, which transfers data from a block referenced by a device pointer (`devptr`) into the given fragment. The `stride` argument identifies the number of bytes between adjacent rows (for fragments with row major layout) or between adjacent columns (for column major layout).

The table has two versions of `load_matrix_sync`, and the second is used for fragments that make up the accumulator matrix (Matrix C). The difference between them is that the second version accepts an argument that identifies the layout of the fragment's data. This can be set to `mem_row_major` for row-major layout and `mem_col_major` for column-major layout.

The `store_matrix_sync` function transfers data from a fragment to device memory. Like the second version of `load_matrix_sync`, this accepts a stride argument that identifies the number of bytes between adjacent rows or columns. The last argument specifies the data's layout, and can be set to `mem_row_major` or `mem_col_major`.

The last function in the table, `mma_sync`, performs the MMA operation with the given fragments. The first four arguments are the fragments in the different matrices. The last argument, `satf`, stands for *saturate to finite value*. If this is set to true, the values in the accumulator will be subject to the following conditions:

- A value at positive infinity will be set to the maximum possible value for the type.
- A value at negative infinity will be set to the minimum possible value for the type.
- If a value is not-a-number (NaN), it will be set to +0.

The names of the last four functions in the table end with `_sync`. This means they behave synchronously, and the application waits for the warp to perform the operation. CUDA provides asynchronous functions for MMA operations as part of its WGMMA (Warp Group Matrix Multiply Accumulate) instructions. These instructions are discussed in Appendix B, which covers Parallel Thread Execution (PTX) in detail.

5.5.3 Example Application

The ch05_wmma_demo project demonstrates how a CUDA application can use the WMMA API to multiply two 128-by-128 matrices and add the product to a third 128-by-128 matrix. The host application performs seven operations:

1. Creates arrays for Matrix A and Matrix B on the host, and initializes them with half-precision values of 1.0.
2. Allocates device blocks for Matrix A and Matrix B on the device, and copies data from Matrix A and Matrix B on the host.
3. Creates an empty array for Matrix C on the host and allocates a block of memory for Matrix C on the device.
4. Configures each thread block to contain 32 threads. This is because each thread block must contain a single warp.
5. Creates a two-dimensional grid with dimensions (8, 8), which means the application relies on a total of 64 thread blocks.
6. Calls the `wmma_demo` kernel with the three pointers to device memory containing Matrix A, Matrix B, and Matrix C.
7. Transfers the output matrix to the host, displays the results, and frees the host memory and device memory containing the matrices.

The `wmma_demo` kernel receives the three pointers to memory and performs six operations:

1. Uses the thread's block ID to determine which tile its warp should process.
2. Creates a fragment for Matrix C and calls `wmma::fill_fragment` to fill it with single-precision zeros.
3. Iterates through the tile dimension and creates a fragment in Matrix A and Matrix B for each tile.
4. For each iteration, it calls `wmma::load_matrix_sync` twice: once to load data for Matrix A and once to load data for Matrix B.
5. For each iteration, it calls `wmma::mma_sync` to multiply the fragments for Matrix A and B and add the values of Matrix C.
6. When the loop is finished, the kernel calls `wmma::store_matrix_sync` to store the results to the block of device memory.

To better understand how these steps are performed, it helps to look at real code. Listing 5.4 presents the code in ch05_wmma_demo/main.cu.

Listing 5.4 ch05_wmma_demo/main.cu

```
__global__ void wmma_demo(const half* A, const half* B, float* C)
{

  // Warp responsible for tile at (tileRow, tileCol)
  int tileRow = blockIdx.y;
  int tileCol = blockIdx.x;

  // Initialize Matrix C (the accumulator)
  wmma::fragment<wmma::accumulator, TILE_M, TILE_N,
    TILE_K, float> frag_c;
  wmma::fill_fragment(frag_c, 0.0f);

  // Loop over the K dimension in tiles
  for (int tileK = 0; tileK < K / TILE_K; tileK++) {

    // Offset pointers
    const half* tileA = A + (tileRow * TILE_M) * K +
      tileK * TILE_K;
    const half* tileB = B + (tileK * TILE_K) * N +
      tileCol * TILE_N;
```

Listing 5.4 ch05_wmma_demo/main.cu (Continued)

```
    // Create fragment A
    wmma::fragment<wmma::matrix_a, TILE_M, TILE_N, TILE_K,
      half, wmma::row_major> frag_a;

    // Create fragment B
    wmma::fragment<wmma::matrix_b, TILE_M, TILE_N, TILE_K,
      half, wmma::col_major> frag_b;

    // Load matrix data
    wmma::load_matrix_sync(frag_a, tileA, K);
    wmma::load_matrix_sync(frag_b, tileB, N);

    // Perform MMA operation
    wmma::mma_sync(frag_c, frag_a, frag_b, frag_c);
  }

  // Store the output C tile
  float* tileC = C + (tileRow * TILE_M) * N + tileCol * TILE_N;
  wmma::store_matrix_sync(tileC, frag_c, N, wmma::mem_row_major);
}

int main() {

  // Host matrix A
  half *hostMatrixA = new half[M * K];
  for (int i = 0; i < M * K; i++) {
    hostMatrixA[i] = __float2half(1.0f);
  }

  // Host matrix B
  half* hostMatrixB = new half[K * N];
  for (int i = 0; i < K * N; i++) {
    hostMatrixB[i] = __float2half(1.0f);
  }

  // Device matrix A
  half* devMatrixA;
  checkCudaErrors(cudaMalloc(&devMatrixA, sizeof(half) * M * K));
  checkCudaErrors(cudaMemcpy(devMatrixA, hostMatrixA,
    sizeof(half) * M * K, cudaMemcpyHostToDevice));
```

Listing 5.4 ch05_wmma_demo/main.cu (Continued)

```
    // Device matrix B
    half* devMatrixB;
    checkCudaErrors(cudaMalloc(&devMatrixB, sizeof(half) * K * N));
    checkCudaErrors(cudaMemcpy(devMatrixB, hostMatrixB,
      sizeof(half) * K * N, cudaMemcpyHostToDevice));

    // Host and device Matrix C
    float* hostMatrixC = new float[M * N];
    float* devMatrixC;
    checkCudaErrors(cudaMalloc(&devMatrixC,
      sizeof(float) * M * N));

    // One warp per thread block
    dim3 blockDim(32, 1, 1);

    // M/16 by N/16 tiles in each matrix
    dim3 gridDim(N / 16, M / 16);

    // Execute kernel
    wmma_demo<<<gridDim, blockDim>>>(devMatrixA,
      devMatrixB, devMatrixC);

    // Wait for kernel to execute
    cudaDeviceSynchronize();

    // Copy data from device to host
    checkCudaErrors(cudaMemcpy(hostMatrixC, devMatrixC,
      sizeof(float) * M * N, cudaMemcpyDeviceToHost));

    // Display results
    std::cout << "C[0] = " << hostMatrixC[0] << std::endl;
    std::cout << "Expected value: 128\n";

    // Free device memory
    checkCudaErrors(cudaFree(devMatrixA));
    checkCudaErrors(cudaFree(devMatrixB));
    checkCudaErrors(cudaFree(devMatrixC));

    // Free host memory
    delete[] hostMatrixA;
    delete[] hostMatrixB;
    delete[] hostMatrixC;
}
```

The dimensions used in this application can be hard to understand. A tile is a block of data that can be processed by a warp, and each tile dimension is 16-by-16. In code, the tile dimensions are given as `TILE_M`, `TILE_N`, and `TILE_K`. The matrices all have the same size of 128-by-128, which corresponds to an 8-by-8 grid of tiles. The matrix dimensions are given as `M`, `N`, and `K`.

Each thread block contains a single warp and each warp contains 32 threads. Each warp processes a tile, and because each matrix is an 8-by-8 grid of tiles, the grid dimension is set to 8-by-8. The following code shows how this is accomplished:

```
// Set dimensions
dim3 blockDim(32, 1, 1);
dim3 gridDim(N / 16, M / 16);

// Execute kernel
wmma_demo<<<gridDim, blockDim>>>(devMatrixA, devMatrixB,
  devMatrixC);
```

The application doesn't create a separate matrix for Matrix D. Instead, the results of the MMA are stored back into Matrix C. This explains why `frag_c` is given twice in the call to `wmma::mma_sync`:

```
wmma::mma_sync(frag_c, frag_a, frag_b, frag_c);
```

The application sets the values in Matrix A and Matrix B to 1.0 and sets the values in Matrix C to 0.0. When the MMA operation is finished, the result should equal the number of values in Matrix A and Matrix B, which is 128.

5.6 Summary

This chapter has presented features of CUDA that aren't absolutely essential, but can improve performance if used correctly. By taking advantage of streams, applications can run device operations in separate queues that execute at the same time. These operations can be synchronized using events, which also make it possible to measure elapsed time.

Normally, the operating system transfers data between RAM and disk, which will cause delays if data needs to be transferred to the GPU. But if an application enables and allocates pinned memory, the data will remain in RAM, allowing the GPU to access it directly. For workloads that move large amounts of data, using pinned memory can significantly improve performance.

Dynamic parallelism makes it possible for one kernel to execute one or more other kernels. This makes it possible to implement recursive algorithms in CUDA, such as tree traversal and graph searching. Unfortunately, this can take a great deal of time, so it should only be used when traditional processing isn't sufficient.

The last part of this chapter shows how the WMMA API can be used to perform matrix-multiply-and-accumulate (MMA) operations at high speed. The main difference between this and other operations is that it combines the processing of the entire warp at once. Appendix B shows how the WGMMA API performs these operations even faster by combining the processing of an entire warp group.

Chapter 6

Accelerating Linear Algebra with cuBLAS

In this chapter, we'll stop discussing general CUDA development and start focusing on Nvidia's libraries. To be specific, we'll explore the cuBLAS library provided in the CUDA Toolkit. The goal is to explain how the functions of this library can be used to perform operations involving vectors and matrices.

Because these operations involve linear transformations, they're collectively referred to as *linear algebra*. Linear algebra forms the basis of many important topics in computer science, including computer graphics and machine learning.

In the late 1970s, researchers recognized the need for a standard set of operations involving vectors and matrices, and they created the Basic Linear Algebra Subprograms (BLAS). BLAS is very important in high-performance computing, and if you look at the specs for a supercomputer, you'll see how quickly it can perform BLAS operations.

BLAS divides linear algebra operations into three groups called *levels*:

- **Level 1** — Operations involving vectors, including dot products and norms
- **Level 2** — Operations involving vectors and matrices
- **Level 3** — Operations involving matrices and matrices

Several implementations of BLAS are available, and the most popular open-source version is OpenBLAS, which can be downloaded from https://github.com/OpenMathLib/OpenBLAS.

This chapter focuses on cuBLAS, which is Nvidia's implementation of BLAS. I'll explain how to write applications that access cuBLAS functions in each of the three levels, and I'll provide example code that shows how the functions are used.

6.1 Overview

One major advantage of using the CUDA Toolkit is simplicity. All the header files are in the include folder and all the library files are in the lib folder. To build a project that uses cuBLAS, you only need to make two changes:

1. In the source code, include the appropriate cuBLAS header file.
2. In the build file, add the cuBLAS library (libcublas.so on Linux, cublas.lib on Windows) to the link procedure.

The first step needs explanation. If you look in the include folder of the CUDA Toolkit, you'll see five header files related to cuBLAS:

- **cublas.h** — contains declarations of deprecated functions (version 1), only used by legacy applications
- **cublas_api.h** — central header file, contains declarations of new and old cuBLAS functions
- **cublas_v2.h** — contains declarations of current (version 2) functions
- **cublasLt.h** — lightweight cuBLAS, contains configurable functions with advanced features
- **cublasXt.h** — multi-GPU cuBLAS, provides functions intended to be run across multiple devices

This chapter focuses on the functions and data structures in cublas_v2.h. Once you understand this, you'll be able to pursue the more advanced features provided in cublasLt.h and cublasXt.h.

6.1.1 General Development

At minimum, the process of coding a cuBLAS application consists of four steps:

1. Initialize a `cublasHandle_t` structure by calling `cublasCreate`.
2. Create data structures to contain input and output data.
3. Call one or more cuBLAS functions to perform linear algebra operations.
4. Free the `cublasHandle_t` structure by calling `cublasDestroy`.

The `cublasHandle_t` structure points to the general cuBLAS context, which contains state information and configuration of the cuBLAS library. This is usually the first argument of the functions provided by cuBLAS.

Just as regular CUDA functions return a `cudaError_t` value, many cuBLAS functions return a `cublasStatus_t` value. This identifies the status of the cuBLAS operation, and the enumerated type has ten values. Table 6.1 lists these values and provides a description of each.

Table 6.1

Status Values

Name	Description
CUBLAS_STATUS_SUCCESS	The operation completed successfully
CUBLAS_STATUS_NOT_INITIALIZED	The cuBLAS library was not initialized properly
CUBLAS_STATUS_ALLOC_FAILED	Failed to allocate cuBLAS resources, usually caused by cudaMalloc
CUBLAS_STATUS_INVALID_VALUE	A value passed to a function was invalid or unsupported
CUBLAS_STATUS_ARCH_MISMATCH	Attempted to access capability unsupported on the target device
CUBLAS_STATUS_MAPPING_ERROR	Failed to access GPU memory space
CUBLAS_STATUS_EXECUTION_FAILED	The program failed to execute
CUBLAS_STATUS_INTERNAL_ERROR	An internal operation failed, usually related to asynchronous operation
CUBLAS_STATUS_NOT_SUPPORTED	The requested functionality is not supported
CUBLAS_STATUS_LICENSE_ERROR	Failed to access license for the desired functionality

For a regular CUDA function, `cudaGetErrorString` accepts the function's return value and returns a string description of the condition. For cuBLAS functions, `cublasGetStatusName` accepts the `cublasStatus_t` value and returns a string description. Its signature is given as follows:

```
const char* cublasGetStatusName(cublasStatus_t status)
```

This function makes it possible to code an error-checking routine that wraps around a cuBLAS function and prints the status description if an error occurs. Throughout this chapter, the following code will be used for this purpose:

```
// Check for cuBLAS errors
inline void checkcuBlasErr(cublasStatus_t err,
  const char* file, int line) {

  if (err != CUBLAS_STATUS_SUCCESS) {
    std::cerr << "cuBLAS error in " << file << " at line "
      << line << ": " << cublasGetStatusName(err) << std::endl;
    std::exit(EXIT_FAILURE);
  }
}
#define checkcuBlasError(err) (checkcuBlasErr((err),
  __FILE__, __LINE__))
```

If `checkcuBlasError` surrounds a cuBLAS function, it will print the status message if an error occurs.

6.1.2 Naming Convention

Every linear algebra function provided by cuBLAS has a name that starts with `cublas`. The rest of the function's name follows the BLAS naming convention, and you can read a full description at https://www.netlib.org/blas.

The name of every cuBLAS function has the same basic structure:

```
cublas<type><op_name>[_64]
```

The `type` portion consists of one or two letters that identify the data type and precision of the elements. This can be set to one of twelve values, and Table 6.2 lists each of them.

Table 6.2

Type Designation Characters

Character(s)	Description
S	Function accepts real single-precision, returns real single-precision
Sc	Function accepts complex single-precision, returns real single-precision
D	Function accepts real double-precision, returns real double-precision
Dz	Function accepts complex double-precision, returns real double-precision

C	Function accepts complex single-precision, returns complex single-precision
Cs	Function accepts real single-precision, returns complex single-precision
Z	Function accepts complex single-precision, returns complex single-precision
Zd	Function accepts real single-precision, returns complex single-precision
Is	Function accepts real single-precision, returns an integer
Id	Function accepts real double-precision, returns an integer
Ic	Function accepts complex single-precision, returns an integer
Iz	Function accepts complex double-precision, returns an integer

Following the `type` designation, `op_name` specifies the operation performed by the function. For example, the function that performs the single-precision dot product is `cublasSdot` and the function that performs the double-precision dot product is `cublasDdot`.

Many cuBLAS functions accept 32-bit integers (`int`s) as arguments. But if an application needs greater precision, 32-bit values may not be sufficient. In this case, cuBLAS provides functions that accept 64-bit integers (`long long`s). If a function accepts these values, its name will end with `_64`.

For example, `cublasIsamax` accepts an array of values and returns the index of the element with the maximum absolute value. The 64-bit version, `cublasIsamax_64`, accepts 64-bit integers as input values and provides a 64-bit integer as the output.

6.1.3 Complex Types

Many cuBLAS functions accept vectors and matrices that contain *complex* elements. Working with these values tends to make programmers nervous, so I'll provide a review:

1. Any number on the real number line, from –infinity to +infinity, is a *real* number.
2. The square root of –1, denoted $\sqrt{-1}$ or i, is an *imaginary* number. The product of a real number (except zero) and an imaginary number is an imaginary number.
3. The sum of a nonzero real number and an imaginary number is a *complex* number.
4. Each complex number, z, has a real part, denoted Re(z), and an imaginary part, denoted Im(z). If z = a + bi, Re(z) = a and Im(z) = b.
5. The conjugate of a complex number, denoted z^*, can be obtained by negating the imaginary part. If z = a + bi, z^* = a – bi.
6. The *modulus* or magnitude of a complex number, denoted |z|, can be obtained by adding the squares of the real and imaginary parts. If z = a + bi, |z| = sqrt($a^2 + b^2$).

7. Complex numbers can be added by adding the real parts and imaginary parts separately. If $z_1 = a + bi$ and $z_2 = c + di$, $z_1 + z_2 = (a + c) + (b + d)i$.
8. Complex numbers can be multiplied in the following manner: if $z_1 = a + bi$ and $z_2 = c + di$, $z_1 * z_2 = (ac - bd) + (ad + bc)i$.

The cuComplex.h header defines two data types that store complex values. `cuComplex` stores the real and imaginary parts as 32-bit `float`s. `cuDoubleComplex` stores the real and imaginary parts as 64-bit `double`s. In essence, a `cuComplex` is a `float2` vector and a `cuDoubleComplex` is a `double2` vector. Chapter 4 discussed these and other vector types in detail.

An application can create a `cuComplex` by calling `make_cuComplex` with the values of the real and imaginary parts. To demonstrate this, the following code creates a complex value with a real part of 3.0 and an imaginary part of 7.0:

```
cuComplex z = make_cuComplex(3.0f, 7.0f);
```

The function that creates a `cuDoubleComplex` is `make_cuDoubleComplex`, and the following code creates a double-precision complex value with the same real and imaginary parts.

```
cuDoubleComplex z = make_cuDoubleComplex(3.0, 7.0);
```

The cuComplex.h header also provides several functions that perform simple operations on complex values. `cuCaddf` adds two `cuComplex` values and returns the sum, `cuCMulf` multiplies two `cuComplex` values, and `cuCdivf` divides two `cuComplex` values.

There are similar functions for double-precision complex values, and their names are obtained by dropping the `f` from the end. That is, `cuCadd` adds two `cuDoubleComplex` values, `cuCMul` multiplies two `cuDoubleComplex` values, and `cuCdiv` divides two `cuDoubleComplex` values.

The cuComplex.h header also provides functions that perform operations beyond arithmetic. These include the following:

- `cuCrealf/cuCreal` — Return the real part of a complex value
- `cuCimagf/cuCimag` — Return the imaginary part of a complex value
- `cuConjf/cuConj` — Return the conjugate of a complex value
- `cuCabsf/cuCabs` — Return the absolute value of a complex value

6.2 First-Level Operations

Chapter 4 introduced the vector types supported by CUDA, such as `int4` and `float3`. The vectors discussed in this chapter are more general, and represent one-dimensional arrays of data. Every element in the vector must have the same type, but the number of elements is unlimited.

BLAS Level 1 functions make it possible to perform operations on these general vectors, and cuBLAS provides implementations of these functions that run on Nvidia devices. Before I describe them, I'd like to briefly review the theory of vectors and their operations.

6.2.1 Understanding Vectors

In graphics and machine learning, one-dimensional quantities like vertex coordinates are frequently represented by vectors. Mathematically, a vector is identified by a lowercase letter in bold, such as **v**. Graphically, a vector is commonly represented by an arrow. To understand BLAS Level 1 functions, you need to be familiar with six points related to vectors:

1. Every vector has a direction. This is graphically represented by the arrow's direction.
2. Every vector has a magnitude. This is graphically represented by the arrow's length.
3. Vectors are defined in a coordinate system. For example, three-dimensional vectors are defined in a coordinate system with three dimensions such as (x, y, z).
4. A vector can be expressed as the combination of n *components*, where n is the number of dimensions in the coordinate system. If **v** is in the (x, y, z) coordinate system, its components are v_x, v_y, and v_z.
5. Vectors can be added by adding their respective components. If **a** and **b** are vectors in the (x, y, z) system and vector **c** = **a** + **b**, the components of **c** equal the sum of the components of **a** and **b**. That is, $c_x = a_x + b_x$, $c_y = a_y + b_y$, and $c_z = a_z + b_z$.
6. The dot product of two vectors can be obtained by multiplying their respective components and adding the products together. If **a** and **b** are vectors in the (x, y, z) system and scalar d = **a** · **b**, d equals the sum of the products of the components of **a** and **b**. That is, $d = a_x b_x + a_y b_y + a_z b_z$.

In BLAS, vectors can have any number of dimensions and any number of components. On the following page, Figure 6.1 shows what a vector looks like in the (x, y) coordinate system.

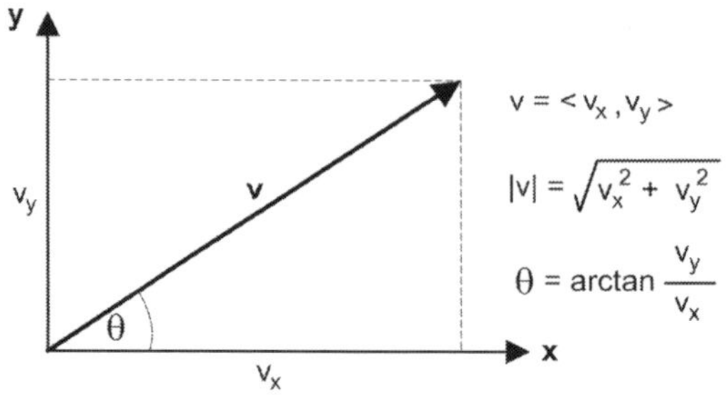

Figure 6.1 A Two-DImensional Vector and Its Components

As illustrated, **v** has components v_x and v_y. v_x is the vector's length along the x-axis and v_y is the vector's length along the y-axis. The total length of **v**, denoted |**v**|, can be computed from v_x and v_y using the Pythagorean theorem.

6.2.2 Creating and Initializing Vectors

Creating a vector to be passed into a cuBLAS function is like creating a regular CUDA array on the device. Two steps are required:

1. Declare a pointer to a variable with the same type as the desired array element.
2. Call `cudaMalloc` with the first argument to the pointer and the second argument set to the number of bytes in the array.

For example, if you'd like to allocate device memory for an array with N `float`s, you can use code like the following:

```
float *x;
cudaMalloc(reinterpret_cast<void**>(&x), N * sizeof(float));
```

You can use similar code to allocate memory for complex-valued arrays. As discussed earlier, the `cuComplex` type represents complex values whose real and imaginary parts are given with single-precision values. The `cuDoubleComplex` type represents complex values whose real and imaginary parts are given with double-precision values.

The following code allocates device memory for an array containing N `cuComplex` values.

```
cuComplex *z;
cudaMalloc(reinterpret_cast<void**>(&z), N * sizeof(cuComplex));
```

Applications can initialize the content of a vector in one of two ways. The first involves calling the `cudaMemcpy` function introduced in Chapter 3. The second is to call the `cublasSetVector` function provided by cuBLAS.

```
cublasSetVector(int n, int elemSize, const void *x, int incx,
  void *y, int incy)
```

This copies `n` values of size `elemSize` from address `x` in host memory to the device memory block referenced by `y`. The main difference between this and `cudaMemcpy` is that you can set the *stride* between vector elements with `incx` and `incy`. This is important to understand:

- `incx` — the number of elements to skip when accessing data referenced by `x`
- `incy` — the number of elements to skip when accessing data referenced by `y`

For example, if `incy` is set to `2`, `cublasSetVector` will access every other value in the `y` array. If `incx` is set to `1`, the function will access every value in x.

Unlike `cudaMemcpy`, `cublasSetVector` can only be used to transfer data from the host to the vector data on the device. To transfer data from the vector data on the device to the host, an application can call `cublasGetVector`:

```
cublasGetVector(int n, int elemSize, const void *x, int incx,
  void *y, int incy)
```

The arguments of the same function have the same types as those of `cublasSetVector`, and serve the same purposes. But in this function, data is transferred from the device memory referenced by `x` to host memory referenced by `y`.

6.2.3 Level 1 Functions

The web site https://www.netlib.org/blas presents the full list of Level 1 BLAS functions. Table 6.3 lists the Level 1 functions provided by cuBLAS, but the signatures don't include the argument types. This is because the data types are identified in the function name using the convention described earlier.

In the first column, each function name contains a question mark that serves as a placeholder for one of the type designations in the second column. For example, cuBLAS provides four forms of `cublas?axpy`: `cublasSaxpy`, `cublasDaxpy`, `cublasCaxpy`, and `cublasZaxpy`.

Table 6.3
Level 1 Functions

Function	Types	Description
`cublas?copy(handle, n, x, incx, y, incy)`	S/D/C/Z	Copy elements of x to elements of y
`cublas?swap(handle, n, x, incx, y, incy)`	S/D/C/Z	Swap elements of x and elements of y
`cublas?scal(handle, n, a, x, incx)`	S/D/C/Z Cs/Zd	Scale elements of x by a
`cublas?axpy(handle, n, a, x, incx, y, incy)`	S/D/C/Z	Scale elements of x by a and add elements of y
`cublas?dot(handle, n, x, incx, y, incy, z)`	S/D	Computes the real-valued dot product of x and y
`cublas?dotc(handle, n, x, incx, y, incy, z)`	C/Z	Computes the conjugated dot product of x and y
`cublas?dotu(handle, n, x, incx, y, incy, z)`	C/Z	Computes the unconjugated dot product of x and y
`cublas?asum(handle, n, x, incx, z)`	S/D/Sc/Dz	Computes the sum of the absolute values of x (1-norm)
`cublas?nrm2(handle, n, x, incx, z)`	S/D/C/Z	Computes the Euclidean sum of the elements in x (2-norm)
`cublas?amax(handle, n, x, incx, z)`	Is/Id/Ic/Iz	Finds the index of the largest element of x
`cublas?amin(handle, n, x, incx, z)`	Is/Id/Ic/Iz	Finds the index of the smallest element of x
`cublas?rot(handle, n, x, incx, y, incy, c, s)`	S/D/C/Z Cs/Zd	Rotates x and y by the angle with the given cosine and sine
`cublas?rotg(handle, x, y, c, s)`	S/D/C/Z	Generate parameters for a standard Givens rotation
`cublas?rotm(handle, n, x, incx, y, incy, param)`	S/D	Apply a modified Givens rotation
`cublas?rotmg(handle, d1, d2, a, b, param)`	S/D	Generate parameters for a modified Givens rotation

These function signatures may look confusing because the arguments don't have data types. But these signatures use BLAS conventions for the variable names, so the types are given as follows:

- `x` and `y` are always vectors (real or complex, single-precision or double-precision).
- `n`, `incx`, and `incy` are always integers — `n` sets the number of elements to be processed, `incx` is the stride of the `x` vector, and `incy` is the stride of the `y` vector.
- The other arguments, such as `a`, `b`, c, and s are floating-point values.

In Table 6.3, the first four functions are the simplest, and `cublas?copy` simply copies vector `x` to vector `y`. `cublas?swap` swaps the elements of the two vectors, setting the elements of `y` equal to the elements of `x` and vice-versa.

The `scal` in `cublas?scal` refers to *scaling*, which is the process of multiplying the elements of a vector or matrix by a scalar. `cublas?scal` scales the vector `x` by the scalar `a`, and `cublas?axpy` scales the vector `x` by the scalar `a` and then adds the vector `y`.

Dot Product

As discussed earlier, the dot product of two vectors is computed by multiplying the corresponding elements together and adding the products. A positive result implies that the vectors point in similar directions, a negative result implies that they point in opposite directions, and a result of zero means that the vectors are perpendicular.

The simplest cuBLAS dot product function is `cublas?dot`, which computes the dot product of two real-valued vectors. If the vectors have complex values, it's common to take the conjugate of the first vector before computing the dot product. This is called the *conjugated* dot product, and is performed by `cublas?dotc`. In contrast, the *unconjugated* dot product is performed by `cublas?dotu`.

The code in the ch06_complex_dot project demonstrates how `cublasCdotc` and `cublasCdotu` are used. The goal is to compute the conjugated and unconjugated dot products of vectors `x` and `y`. Both vectors contain three complex numbers, and Figure 6.2 shows what they look like.

x =	1 + 2i	3 + 4i	5 + 6i
y =	6 + 5i	4 + 3i	2 + i

cublasCdotc(x, y) $= (1 - 2i)(6 + 5i) + (3 - 4i)(4 + 3i) + (5 - 6i)(2 + i) = 56 - 21i$

cublasCdotu(x, y) $= (1 + 2i)(6 + 5i) + (3 + 4i)(4 + 3i) + (5 + 6i)(2 + i) = 59i$

Figure 6.2 Example Complex-Valued Dot Product

As illustrated, the two dot products produce different results because the first function takes the conjugate of x and the second function doesn't. Listing 6.1 presents the code in ch06_complex_dot/main.cu, which calls both functions and displays their results.

Listing 6.1 ch06_complex_dot/main.cu

```
int main() {
  // Allocate space for vectors
  cuFloatComplex *xMem, *yMem;
  checkCudaErrors(cudaMalloc(reinterpret_cast<void**>(&xMem),
    N * sizeof(cuFloatComplex)));
  checkCudaErrors(cudaMalloc(reinterpret_cast<void**>(&yMem),
    N * sizeof(cuFloatComplex)));

  // Transfer data to vectors
  float xData[] = {1.0f, 2.0f, 3.0f, 4.0f, 5.0f, 6.0f};
  float yData[] = {6.0f, 5.0f, 4.0f, 3.0f, 2.0f, 1.0f};
  checkcuBlasError(cublasSetVector(N, sizeof(cuFloatComplex),
    xData, 1, xMem, 1));
  checkcuBlasError(cublasSetVector(N, sizeof(cuFloatComplex),
    yData, 1, yMem, 1));

  // Create cuBLAS handle
  cublasHandle_t handle;
  checkcuBlasError(cublasCreate(&handle));

  // Perform computations
  cuFloatComplex z1, z2;
  checkcuBlasError(cublasCdotc(handle, N, xMem,
    1, yMem, 1, &z1));
  checkcuBlasError(cublasCdotu(handle, N, xMem,
    1, yMem, 1, &z2));

  // Display results
  std::cout << "Conjugated dot product = " << cuCrealf(z1)
    << ", " << cuCimagf(z1) << "i" << std::endl;
  std::cout << "Unconjugated dot product = " << cuCrealf(z2)
    << ", " << cuCimagf(z2) << "i" << std::endl;

  // Free resources
  checkcuBlasError(cublasDestroy(handle));
  cudaFree(xMem);
  cudaFree(yMem);
  return 0;
}
```

If you compare this code to examples from previous chapters, the first thing to notice is that there's no kernel. The `cublasCdotc` and `cublasCdotu` functions access the device directly, so there's no need to write separate routines. When these functions return, they provide the result of the computation.

The pointers to device memory, `xMem` and `yMem`, have type `cuFloatComplex` because the dot product functions require complex values. However, the initialization data variables, `xData` and `yData`, are regular `float` arrays. When you initialize an array of `cuFloatComplex` or `cuDoubleComplex` values, you can provide the `float`s or `double`s in a regular array as long as the values are provided in real-imaginary pairs.

Once the allocation and initialization steps are performed, the application calls `cublasCreate` to obtain a handle that represents the execution environment. Then it calls the two dot product functions, displays their results, and frees resources, including the handle.

cuBLAS doesn't provide a simple way to print complex values to standard output. For this reason, the application calls `cuCrealf` to obtain the real part of each complex value and `cuCimagf` to obtain the imaginary part.

Norms

A norm is a function that accepts a vector and returns a real value greater than or equal to zero. If **x** is a vector, its norm can be denoted ||**x**||. A p-norm, denoted $||x||_p$ is a norm that can be expressed with the following equation:

$$\| x \|_p = \left(\sum_{i=0}^{n-1} | x_i |^p \right)^{\frac{1}{p}}$$

The next three functions in the table compute p-norms for different values of p. The first, `cublas?asum`, computes the 1-norm, which is simply the sum of the absolute value of the vector's elements. The next function, `cublas?nrm2`, computes the 2-norm, more commonly called the *Euclidean sum*. The equation is given as follows:

$$\| x \|_2 = \left(\sum_{i=0}^{n-1} | x_i |^2 \right)^{\frac{1}{2}}$$

When p reaches infinity, the p-norm simply equals the largest value of the vector. The third function, `cublas?amax`, doesn't return the maximum, but instead returns the index of the first occurrence of the maximum value. Similarly, the following function, `cublas?amin`, returns the index of the minimum value in a vector.

Rotation

The last four functions in the table are concerned with rotating vectors in space, and have nothing to do with bitwise rotation. The first function, `cublas?rot`, rotates **x** and **y** by an angle, but it doesn't accept the angle as a parameter. Instead, the last two values identify the angle's cosine and sine.

The `cublas?rot` function doesn't return its results in separate vectors. It's an *in-place* function, which means it overwrites the inputs x and y. If `c` is the cosine of the angle and `s` is the sine of the angle, then the function updates x and y in the following way:

$$x' = cx + sy$$
$$y' = -sx + cy$$

Many linear algebra operations, such as QR decomposition, require a specific rotation that sets components of the output vector to zero. This is called a *Givens rotation*, and referring to the preceding equations, the goal is to find values of c and s that set y' to zero.

This computation is performed by `cublas?rotg`. It doesn't perform rotation, but instead computes the values of the cosine and sine needed for the Givens rotation. The last two parameters, `c` and `s`, are the cosine/sine values that make the transformed y vector equal to 0.

Givens rotations are useful, but if the input values are very large or very small, errors can result due to overflow or underflow. For this reason, cuBLAS also makes it possible to rotate vector using *modified Givens rotations*, which improve stability by performing scaling in addition to rotation.

An application can perform these modified Givens rotations by calling the `cublas?rotm` function. The last argument needs to be set to an array containing four parameters that are commonly called h_{11}, h_{12}, h_{21}, and h_{22}. The equation for the modified rotation is given as follows:

$$x' = h_{11}x + h_{12}y$$
$$y' = h_{21}x + h_{22}y$$

The equations needed to compute h_{11}, h_{12}, h_{21}, and h_{22} are complicated, but there's no need to determine them directly. Applications can obtain them by calling `cublas?rotmg`, which returns the four parameters in its last argument.

6.3 Second-Level Operations

The operations in BLAS Layer 2 combine vectors and matrices. Most of these functions involve multiplying a matrix by a vector, but there are a handful of functions that solve systems of equations.

This section presents the cuBLAS functions that perform BLAS Level 2 operations. Before I delve into the code, I'd like to provide a brief overview of matrix theory.

6.3.1 Understanding Matrices

Simply put, a matrix is a rectangular arrangement of numbers. Symbolically, matrices are denoted with capital letters in boldface, such as **A**. Graphically, a matrix is represented as a grid of numbers inside vertical bars. This is shown in Figure 6.3.

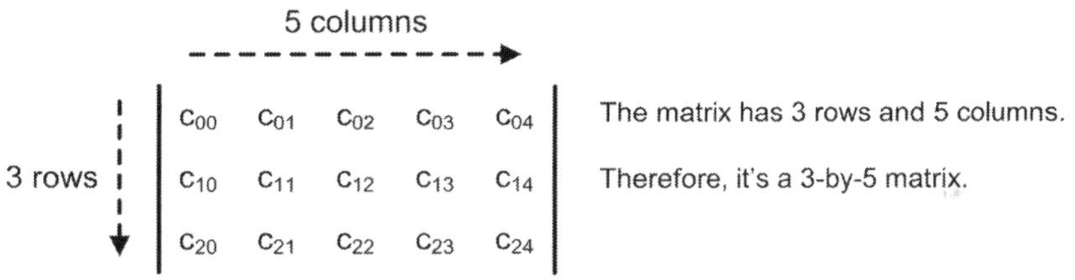

Figure 6.3 Matrix Notation

The numbers inside a matrix are arranged in rows and columns. If a matrix has m rows and n columns, it's called an m-by-n matrix. Each row and column is a one-dimensional structure of numbers, and for this reason, each row and column can be thought of as a vector.

The numbers inside a matrix are *elements*. In Figure 6.3, the element c_{ij} belongs to the ith row and the jth column. If i equals j, then the element lies on an imaginary line called the matrix's *main diagonal*, which runs from the upper-left to the lower-right.

Classifying Matrices

Mathematicians have devised many ways to classify matrices, and some cuBLAS functions require matrices with specific features. If the elements on the matrix diagonal are nonzero and elements off the diagonal are zero, the matrix is a *diagonal matrix*. If the nonzero elements of a diagonal matrix all equal 1, the matrix is an *identity matrix*.

If the elements above the diagonal, c_{ij}, equal the elements below the main diagonal, c_{ji}, the matrix is *symmetric*. If all the elements above the diagonal are zero, the matrix is *lower triangular*. If all the elements below the diagonal are zero, the matrix is *upper triangular*.

Diagonals above the main diagonal are called *superdiagonals* and diagonals below the main diagonal are called *subdiagonals*. A matrix is *banded* if its nonzero elements can only be found on the main diagonal or on a limited number of superdiagonals and subdiagonals.

Figure 6.4 shows what a banded 6-by-6 matrix looks like. There are nonzero elements on the main diagonal, the diagonal immediately above the main diagonal, and the diagonal immediately below the main diagonal. All the other elements are zero. Therefore, this banded matrix has one superdiagonal and one subdiagonal.

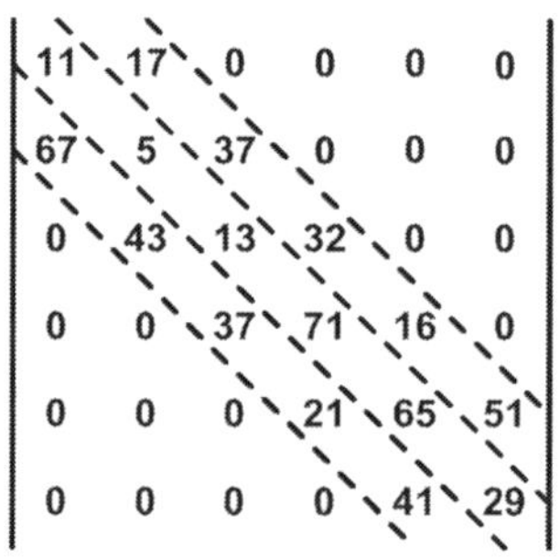

Figure 6.4 A Banded Matrix

Matrices can have real or complex values. If a matrix contains complex values and each element above the main diagonal is the conjugate of the corresponding element below the matrix, the matrix is *Hermitian*. Put another way, a matrix is Hermitian if c_{ij} equals c_{ji}^* for all i and j.

If the overwhelming majority of elements in a matrix are zero, operations can be employed that only take the nonzero elements into account. In this case, the matrix is called a *sparse matrix*. Matrices that aren't sparse are called *dense matrices*. Chapter 7 will discuss dense and sparse matrices in detail.

Matrix-Vector Multiplication

In graphics and robotics, applications frequently need to transform points in space to other points in space. For example, if you zoom in on a three-dimensional rendering of a figure, the application will need to transform every vertex of the figure to make it appear larger.

Mathematically, these and other operations are accomplished using a process called *matrix-vector multiplication*, where the matrix identifies the transformation and the vector contains the coordinates of a vertex. To understand how this works, it helps to think of each row of a matrix as a vector. When a vector is multiplied by a matrix, the resulting vector contains the dot products of the vector and each row of the matrix.

Figure 6.5 depicts the process of multiplying a 4-by-4 matrix (**A**) with a vector (**x**) containing four elements. Each row of the matrix is circled to emphasize that it's a separate vector. The dot product of each row of **A** with **x** produces a value in the output vector **b**.

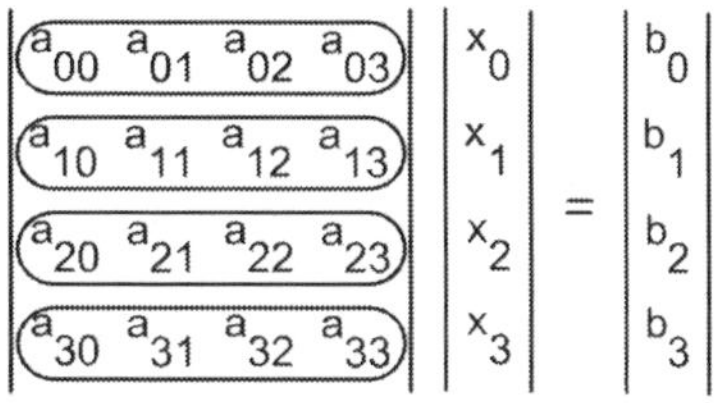

Figure 6.5 Matrix-Vector Multiplication

In this example, multiplication requires taking the dot product of each row of **A** and the **x** vector. Because there are four rows, there are four dot products, and the results are placed in the **b** vector. The following equations should help make this clear:

- $b_0 = a_{00}x_0 + a_{01}x_1 + a_{02}x_2 + a_{03}x_3$
- $b_1 = a_{10}x_0 + a_{11}x_1 + a_{12}x_2 + a_{13}x_3$
- $b_2 = a_{20}x_0 + a_{21}x_1 + a_{22}x_2 + a_{23}x_3$
- $b_3 = a_{30}x_0 + a_{31}x_1 + a_{32}x_2 + a_{33}x_3$

A vector can only be multiplied by a matrix if each row of the matrix has the same number of elements as the vector. If an M-by-N matrix multiplies a vector with N elements, M dot products will be necessary and the result will have M elements.

Packed Matrices

In practical applications, a matrix may contain tens or hundreds of thousands of elements. If a matrix is triangular, symmetric, or Hermitian, it isn't necessary to store every element. Instead, the application only needs to provide values in the upper or lower triangle. These matrices are called packed matrices.

BLAS and cuBLAS support operations involving packed matrices. Each function needs to identify if the data represents the matrix's upper or lower triangle.

6.3.2 Creating and Initializing Matrices

Working with cuBLAS matrices is like working with vectors. There are three main steps:

1. Allocate a block of device memory for the matrix with `cudaMalloc`.
2. Transfer data from the host to the matrix with `cudaMemcpy` or `cublasSetMatrix`.
3. Transfer data from the matrix to the host with `cudaMemcpy` or `cublasGetMatrix`.

If you understood the Level 1 BLAS example code, the functions used for Level 2 won't present any difficulty. But there's one important point to keep in mind: cuBLAS stores and accesses matrix elements in *column-major order*. That is, cuBLAS stores elements in the same *column* next to one another in memory.

To understand this, it helps to look at an example. Suppose you want to initialize a matrix so that it takes the form shown in Figure 6.6.

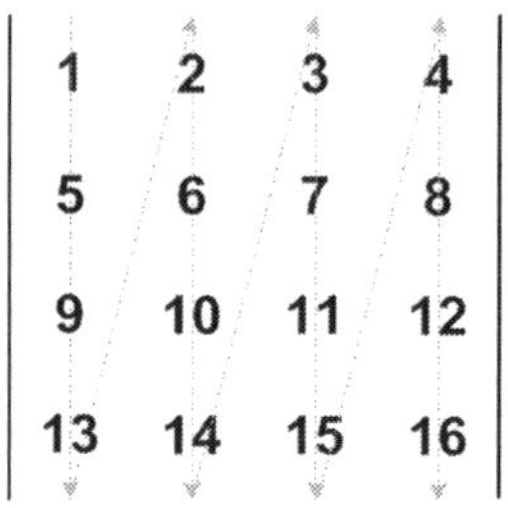

Figure 6.6 Matrix Elements in Column-Major Order

To initialize the matrix, the host needs to provide the elements in the order indicated by the dashed gray arrows. That is, the first value is 1, followed by 5, 9, and 13. The next four values are 3, 7, 11, and 15. The following code shows how the array containing the matrix data could be defined using `float`s.

```
float matrixData[] = {1.0f, 5.0f, 9.0f, 13.0f, 2.0f, 6.0f, 10.0f,
  14.0f, 3.0f, 7.0f, 11.0f, 15.0f, 4.0f, 8.0f, 12.0f, 16.0f};
```

When this array is transferred to the device, the first four values will be set to the first column of the matrix, and then the second column will be set, and so on. The next example project will demonstrate how this works.

Applications can write data to a matrix by calling `cudaMemcpy`. If the columns of the input or output matrix have extra bytes at the end (padding), it's better to call `cublasSetMatrix`, whose signature is given as follows:

```
cublasSetMatrix(int rows, int cols, int elemSize, const void *A,
  int lda, void *B, int ldb)
```

The first two arguments identify the number of rows and columns, and the third sets the size of each element. The fifth argument, A, points to the matrix data on the host, and the seventh argument, B, points to the block of memory that will contain the matrix on the device.

The sixth and eighth arguments deserve explanation. `lda` stands for *leading dimension of A*, and identifies the number of elements between consecutive columns in the host data. In a normal matrix, this simply equals the number of rows. But if each column of the host memory ends with extra data, `lda` should be set to a value greater than the number of rows.

Similarly, `ldb` stands for *leading dimension of B*, and identifies the number of values between consecutive columns in device memory. Throughout this chapter, `lda` and `ldb` will be set to the number of rows, which means there's no padding at the end of columns on the host or device.

Just as `cublasSetMatrix` initializes a matrix on the device with data from the host, `cublasGetMatrix` reads matrix data from the device and stores it on the host. Its signature is nearly identical to that of `cublasSetMatrix`:

```
cublasGetMatrix(int rows, int cols, int elemSize, const void *A,
  int lda, void *B, int ldb)
```

The main difference between the two functions involves the pointers. In `cublasGetMatrix`, the first pointer points to device memory and the second pointer references host memory.

6.3.3 Level 2 Functions

The Level 2 BLAS routines are essentially similar to the Level 1 routines, but instead of multiplying vectors, most of them perform matrix-vector multiplication. Each matrix requires a leading dimension value (`lda` or `ldb`) and each vector requires a stride value (`incx` or `incy`).

Table 6.4 lists the Level 2 functions provided by cuBLAS. Each accepts a `cublasHandle_t` structure as its first argument, and this can be created by calling `cublasCreate`. In the first column, each function name has a question mark that serves as a placeholder for one of the type designations in the second column. `S` refers to real single-precision, `D` refers to real double-precision, `C` refers to complex single-precision, and `Z` refers to complex double-precision.

Table 6.4

Level 2 Functions

Function	Types	Description
`cublas?gemv(handle, op, m, n, *alpha, *A, lda, *x, incx, *beta, *y, incy)`	S/D/C/Z	General matrix-vector multiplication with addition (y = alpha A* x + beta y)
`cublas?gbmv(handle, op, m, n, kl, ku, *alpha, *A, lda, *x, incx, *beta, *y, incy)`	S/D/C/Z	Banded matrix-vector multiplication with addition (y = alpha A* x + beta y)
`cublas?hemv(handle, ul, m, n, *alpha, *A, lda, *x, incx, *beta, *y, incy)`	S/D/C/Z	Hermitian matrix-vector multiplication with addition (y = alpha A x + beta y)
`cublas?hbmv(handle, ul, m, n, *alpha, *A, lda, *x, incx, *beta, *y, incy)`	S/D/C/Z	Hermitian matrix-vector multiplication with addition (y = alpha A x + beta y)
`cublas?hpmv(handle, ul, m, n, *alpha, *Ap, *x, incx, *beta, *y, incy)`	S/D/C/Z	Packed Hermitian matrix-vector multiplication with addition (y = alpha A x + beta y)
`cublas?symv(handle, ul, m, n, *alpha, *A, lda, *x, incx, *beta, *y, incy)`	S/D/C/Z	Symmetric matrix-vector multiplication with addition (y = alpha A x + beta y)
`cublas?sbmv(handle, ul, m, n, *alpha, *A, lda, *x, incx, *beta, *y, incy)`	S/D/C/Z	Symmetric matrix-vector multiplication with addition (y = alpha A x + beta y)
`cublas?spmv(handle, ul, m, n, *alpha, *Ap, *x, incx, *beta, *y, incy)`	S/D/C/Z	Packed symmetric matrix-vector multiplication with addition (y = alpha A x + beta y)
`cublas?ger(handle, m, n, *alpha, *x, incx, *y, incy, *A, lda)`	S/D	Rank-1 update for a general matrix (A = A + alpha x y^T)
`cublas?geru(handle, m, n, *alpha, *x, incx, *y, incy, *A, lda)`	C/Z	Complex rank-1 update for a general matrix (A = A + alpha x y^T)
`cublas?gerc(handle, m, n, *alpha, *x, incx, *y, incy, *A, lda)`	C/Z	Complex conjugate rank-1 update for a general matrix (A = A + alpha x y^H)

`cublas?syr(handle, ul, n, *alpha, *x, incx, *A, lda)`	S/D	Symmetric rank-1 update (A = A + alpha x x^T)
`cublas?spr(handle, ul, n, *alpha, *x, incx, *Ap)`	S/D	Packed symmetric rank-1 update (A = A + alpha x x^T)
`cublas?her(handle, ul, n, *alpha, *x, incx, *A, lda)`	C/Z	Hermitian rank-1 update (A = A + alpha x x^H)
`cublas?hpr(handle, ul, n, *alpha, *x, incx, *Ap)`	C/Z	Packed Hermitian rank-1 update (A = A + alpha x x^H)
`cublas?syr2(handle, ul, n, *alpha, *x, incx, *y, incy, *A, lda)`	S/D	Symmetric rank-2 update (A = A + alpha x y^T + alpha y x^T)
`cublas?spr2(handle, ul, n, *alpha, *x, incx, *y, incy, *Ap)`	S/D	Packed symmetric rank-2 update (A = A + alpha x y^T + alpha y x^T)
`cublas?her2(handle, ul, n, *alpha, *x, incx, *y, incy, *A, lda)`	C/Z	Hermitian rank-2 update (A = A + alpha x y^T + alpha y x^T)
`cublas?hpr2(handle, ul, n, *alpha, *x, incx, *y, incy, *Ap)`	C/Z	Packed Hermitian rank-2 update (A = A + alpha x y^T + alpha y x^T)
`cublas?trmv(handle, ul, op, di, n, *A, lda, *x, incx)`	S/D/C/Z	Triangular matrix-vector multiplication (x = A* x)
`cublas?tbmv(handle, ul, op, di, n, k, *A, lda, *x, incx)`	S/D/C/Z	Banded triangular matrix-vector multiplication (x = A* x)
`cublas?tpmv(handle, ul, op, di, n, *Ap, *x, incx)`	S/D/C/Z	Packed triangular matrix-vector multiplication (x = A* x)
`cublas?trsv(handle, ul, op, di, n, *A, lda, *x, incx)`	S/D/C/Z	Solve triangular matrix system
`cublas?tbsv(handle, ul, op, di, n, k, *A, lda, *x, incx)`	S/D/C/Z	Solve banded triangular matrix system
`cublas?tpsv(handle, ul, op, di, n, *Ap, *x, incx)`	S/D/C/Z	Solve packed triangular matrix system

The function names may seem confusing at first, but in each case, the first letter after the question mark identifies the type of the matrix required by the function. General matrices are identified by `g`, banded matrices by `b`, symmetric matrices by `s`, Hermitian matrices by `h`, and triangular matrices by `t`.

The first eight functions end with `mv`, which means they perform matrix-vector multiplication. To be precise, they multiply the `A` matrix by the `x` vector, scale the result by `alpha`, and then add the scaled result to the vector `y` scaled by `beta`. Most of the function arguments are similar to those used in Level 1 functions, but the second argument requires explanation.

For `cublas?gemv` and `cublas?gbmv`, the second argument is `op`, which identifies the operation to be performed on the matrix before it's used in multiplication. This can be set to one of three values:

- `CUBLAS_OP_N` — No operation
- `CUBLAS_OP_T` — Transpose
- `CUBLAS_OP_C` — Conjugation

Many of the functions involving symmetric, Hermitian, and triangular matrices have the second argument set to `ul`, which identifies which part of the matrix contains nonzero values. This can be set to one of three values:

- `CUBLAS_FILL_MODE_LOWER` — The lower triangle contains nonzero values
- `CUBLAS_FILL_MODE_UPPER` — The upper triangle contains nonzero values
- `CUBLAS_FILL_MODE_FULL` — The upper triangle and lower triangle contain nonzero values

If the second letter after the question mark is `b`, as in `cublas?hbmv`, the function requires a banded matrix. As discussed earlier, a banded matrix only contains nonzero elements on the main diagonal and in a limited number of superdiagonals and subdiagonals. To specify matrix structure, these functions require additional arguments named `ku` and `kl`. `ku` identifies the number of superdiagonals containing nonzero elements and `kl` identifies the number of subdiagonals containing nonzero elements. The function will use this information to accelerate the linear algebra computation.

If the second letter after the question mark is `p`, as in `cublas?spmv`, the function requires a packed matrix. Instead of requiring `A` and `lda`, these functions require the matrix data to be provided using the packed format. This consists of a one-dimensional array containing elements of the upper triangle or lower triangle (identified with the `ul` argument).

An example will clarify how matrix packing works. Suppose a 4-by-4 matrix is lower triangular, which means that it's square and all the elements above the main diagonal equal zero. The first element in the packed array is c_{00}, followed by the rest of the elements in the first column. The next element is c_{11}, followed by the elements in the column below c_{11}. This is followed by c_{22} and the element below c_{22}. The last element in the packed array is c_{33}.

Example Matrix-Vector Multiplication

The ch06_matrix_vec project demonstrates how matrix-vector multiplication can be implemented using cuBLAS. It calls `cublasSgemv` to multiply the 4-by-4 matrix `A` by the four-element vector `x`. The result is stored in the vector `y`. Figure 6.7 shows what this operation looks like.

$$\begin{vmatrix} 1 & 3 & 5 & 7 \\ 2 & 6 & 10 & 14 \\ 4 & 9 & 15 & 21 \\ 6 & 12 & 20 & 28 \end{vmatrix} \begin{vmatrix} 3 \\ 7 \\ 11 \\ 17 \end{vmatrix} = \begin{vmatrix} 198 \\ 396 \\ 597 \\ 798 \end{vmatrix}$$

Figure 6.7 Example Matrix-Vector Multiplication

If you look through the code in ch06_matrix_vec/main.cu, you'll see that it performs eight operations:

1. Calls `cudaMalloc` to allocate device memory for **A**, **x**, and **y**.
2. Defines the host arrays containing the data of **A** and **x**.
3. Calls `cudaMemcpy` to initialize **A** and **x**.
4. Calls `cublasCreate` to obtain a handle (`cublasHandle_t`).
5. Calls `cublasSgemv` to perform the matrix-vector multiplication.
6. Calls `cudaMemcpy` to transfer **y** to the host.
7. Displays the elements of y.
8. Frees the handle and blocks of device memory.

It's important to note that `cublasSgemv` does more than just multiply A and x. It scales **A** by alpha before performing the multiplication and then adds the product to **y** scaled by beta. The operation's result is stored in the **y** vector.

In the interests of simplicity, alpha and beta are set to 1.0 and **y** is initially left to contain zeros. In addition, the stride of **x** (`incx`) is set to 1.0, the stride of **y** (`incy`) is set to 1.0, and the leading dimension of A (`lda`) is set to the number of rows (`m`).

Listing 6.2 presents the code in ch06_matrix_vec/main.cu. The second argument of `cublasSgemv` is set to `CUBLAS_OP_N`, so no operation is performed on the matrix before the multiplication.

Listing 6.2 ch06_matrix_vec/main.cu

```
int main() {

  // Set matrix dimensions
  const int m = 4;  // Number of rows
  const int n = 4;  // Number of columns

  // Allocate device memory
  float *aMem, *xMem, *yMem;
  checkCudaErrors(cudaMalloc(reinterpret_cast<void**>(&aMem),
    m * n * sizeof(float)));
  checkCudaErrors(cudaMalloc(reinterpret_cast<void**>(&xMem),
    n * sizeof(float)));
  checkCudaErrors(cudaMalloc(reinterpret_cast<void**>(&yMem),
    m * sizeof(float)));

  // Define the A matrix data
  float aData[m * n] = {
    1.0f, 2.0f, 4.0f, 6.0f,      // Col 0
    3.0f, 6.0f, 9.0f, 12.0f,     // Col 1
    5.0f, 10.0f, 15.0f, 20.0f,   // Col 2
    7.0f, 14.0f, 21.0f, 28.0f,   // Col 3
  };

  // Set the x vector data
  float xData[n] = { 3.0f, 7.0f, 11.0f, 17.0f };

  // Transfer data to device
  checkCudaErrors(cudaMemcpy(aMem, aData, m*n*sizeof(float),
    cudaMemcpyHostToDevice));
  checkCudaErrors(cudaMemcpy(xMem, xData, n*sizeof(float),
    cudaMemcpyHostToDevice));

  // Create cuBLAS handle
  cublasHandle_t handle;
  checkcuBlasError(cublasCreate(&handle));

  // Perform y = alpha * A * x + beta * y
  // A is m x n, so lda = m
  float alpha = 1.0f, beta = 1.0f;
```

Listing 6.2 ch06_matrix_vec/main.cu (Continued)

```
  checkcuBlasError(cublasSgemv(
    handle,
    CUBLAS_OP_N,      // No operation
    m,                // Number of rows
    n,                // Number of columns
    &alpha,           // Scaling factor for A
    aMem,             // A matrix
    m,                // Leading dimension of A
    xMem,             // x vector
    1,                // Stride for x
    &beta,            // Scaling factor for x
    yMem,             // y vector
    1                 // Stride for y
  ));

  // Copy y vector to host
  float yData[m];
  checkCudaErrors(cudaMemcpy(yData, yMem, m * sizeof(float),
    cudaMemcpyDeviceToHost));

  // Print result
  std::cout << "Result y = [ ";
  for (int i = 0; i < m; ++i) {
    std::cout << yData[i] << " ";
  }
  std::cout << "]" << std::endl;

  // Destroy handle
  checkcuBlasError(cublasDestroy(handle));

  // Free resources
  cudaFree(aMem);
  cudaFree(xMem);
  cudaFree(yMem);
  return 0;
}
```

This code processes single-precision real values, so there's no need to be concerned about conjugate values. But if the matrix contains complex values, `cublas?gemv` can take the conjugate of **A** before performing multiplication depending on the second argument. This is important to keep in mind when dealing with complex-valued data.

Rank Updates

After the matrix-vector multiplication functions, the next eleven functions in the table perform *rank updates*. Before I can describe these operations, I need to explain what the term *rank* means in linear algebra.

A set of vectors is *linearly independent* if none of the vectors can be obtained by adding and scaling other vectors in the set. The rank of a matrix is the maximum number of its rows or columns that are linearly independent. For example, if the rows of an N-by-N matrix form a linearly independent set, the matrix has rank N.

In BLAS terminology, a rank update is an operation that changes a matrix by adding a matrix of lower rank to it. A Rank-1 update adds a matrix of rank 1 and a Rank-2 update adds a matrix of rank 2.

The rank update functions accept two vectors, **x** and **y**, and form a matrix by taking their *outer product*. Each element of the matrix is obtained by multiplying an element of x by an element of **y**. To be precise, if the outer product of x and y form matrix **C**, then each element c_{ij} equals x_i times y_j.

The outer product of **x** and **y** is denoted as $\mathbf{xy}^T$. Therefore, the update of matrix **A** performed by `cublas?ger` can be expressed with the equation $\mathbf{A} = \mathbf{A} + \alpha\mathbf{xy}^T$. This is called a Rank-1 update because the matrix $\mathbf{xy}^T$ has a rank of at least 1.

If A is symmetric, the Rank-1 update takes the outer product of **x** with itself, so the `cublas?syr` function performs the operation $\mathbf{A} = \mathbf{A} + \alpha\mathbf{xx}^T$. If **A** has complex values, the function `cublas?gerc` takes the conjugate of $\mathbf{xy}^T$, denoted $\mathbf{xy}^H$. In this case, the equation can be expressed as $\mathbf{A} = \mathbf{A} + \alpha\mathbf{xy}^H$.

If the outer product of **x** and **y** is added to the outer product of **y** and **x**, the result is a matrix with a rank of at most 2. Therefore, the functions that perform Rank-2 updates, such as `cublas?syr2` and `cublas?her2` perform the operation represented by the equation $\mathbf{A} = \mathbf{A} + \alpha\mathbf{xy}^T + \alpha\mathbf{yx}^T$.

Triangular Matrix Operations

The last six functions in the table have a `t` after the question mark because they perform operations involving triangular matrices. These functions are different than the others because of their second, third, and fourth parameters:

- `ul` — Identifies if the matrix is upper or lower triangular
- `op` — Sets the operation to be performed on the matrix before the main operation
- `di` — Identifies whether the main diagonal consists of unit values (ones) that shouldn't be changed

The fourth argument, `di`, requires explanation. If all the elements on the main diagonal of a triangular matrix equal 1, it's a *unit triangular matrix*. In some cases, applications may not want these values to change, and this behavior can be configured by setting `di` to one of two values:

- `CUBLAS_DIAG_NON_UNIT` — the main diagonal contains values other than 1
- `CUBLAS_DIAG_UNIT` — all the values on the main diagonal equal 1

For the last three functions, the triangular matrix represents a series of equations and the goal of each function is to find the values that solve these equations. These functions won't be discussed here, but Chapter 7 will provide a detailed explanation of how cuBLAS can be used to solve systems of equations.

6.4 Third-Level Operations

The third level of BLAS routines perform operations related to matrix-matrix multiplication. In addition, cuBLAS provides functions that perform rank updates and system solution. But before I present these functions, I need to explain how matrices can be multiplied together.

6.4.1 Matrix-Matrix Multiplication

As mentioned earlier, matrix-vector multiplication is used to apply a transformation. Similarly, matrix-matrix multiplication is used to apply multiple transformations at the same time. This makes it important in computer graphics and robotics. It's also needed for many system solution algorithms, which makes it necessary in fields like engineering and economics.

Like matrix-vector multiplication, matrix-matrix multiplication consists of a series of dot products. If matrix **C** equals matrix **A** multiplied by matrix **B**, then every element of **C** is obtained by taking the dot product of a row of **A** and a column of **B**.

On the following page, Figure 6.8 illustrates how one of the elements of **C** can be computed. Row 2 has been circled in the **A** matrix and Column 1 has been circled in the **B** matrix. Taking the dot product of this row and column produces the element c_{21} in the C matrix.

Using i as the row index and j as the column index, we can obtain a more general relationship for matrix-matrix multiplication. If matrix **C** equals matrix **A** multiplied by matrix **B**, then c_{ij} equals the dot product of ith row of **A** and the jth column of **B**.

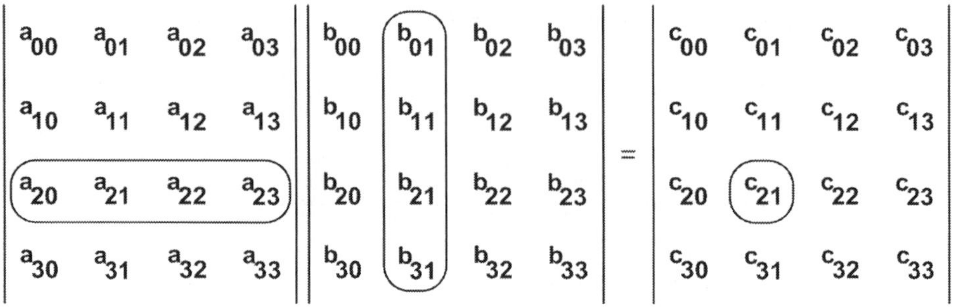

Figure 6.8 Matrix-Matrix Multiplication

If matrix **A** has dimensions P-by-M and matrix **B** has dimensions M-by-Q, **C** will have dimensions P-by-Q. This means the multiplication will require PQ dot products.

6.4.2 Level 3 Functions

Table 6.5 lists the cuBLAS functions that perform BLAS Level 3 operations. The question mark is a placeholder for the elements' type (real/complex) and precision (single/double). This can be replaced with any character in the second column:

- `S` — real-valued single-precision floating-point values
- `D` — real-valued double-precision floating-point values
- `C` — complex-valued single-precision floating-point values
- `Z` — complex-valued double-precision floating-point values

The first three functions perform matrix-matrix multiplication, the next four perform rank updates, and the last two functions deal with triangular matrices.

Table 6.5

Level 3 Functions

Function	Types	Description
`cublas?gemm(handle, opA, opB, m, n, k, *alpha, *A, lda, *B, ldb, *beta, *C, ldc)`	`S/D/C/Z`	General matrix-matrix multiplication with addition (C = alpha AB + beta C)
`cublas?symm(handle, side, ul, m, n, *alpha, *A, lda, *B, ldb, *beta, *C, ldc)`	`S/D/C/Z`	Symmetric matrix-matrix multiplication with addition (C = alpha AB + beta C)

cublas?hemm(handle, side, ul, m, n, *alpha, *A, lda, *B, ldb, *beta, *C, ldc)	C/Z	Hermitian matrix-matrix multiplication with addition (C = alpha AB + beta C)
cublas?syrk(handle, ul, op, n, k, *alpha, *A, lda, *beta, *C, ldc)	S/D/C/Z	Rank-k update of symmetric C (C = alpha AA^T + beta C)
cublas?herk(handle, ul, op, n, k, *alpha, *A, lda, *beta, *C, ldc)	C/Z	Rank-k update of Hermitian C (C = alpha AA^T + beta C)
cublas?syr2k(handle, ul, op, n, k, *alpha, *A, lda, *B, ldb, *beta, *C, ldc)	S/D/C/Z	Rank-2k update of symmetric C (C = alpha AB^T + alpha BA^T + beta C)
cublas?her2k(handle, ul, op, n, k, *alpha, *A, lda, *B, ldb, *beta, *C, ldc)	C/Z	Rank-2k update of Hermitian C (C = alpha AB^H + alpha BA^H + beta C)
cublas?trmm(handle, side, ul, op, di, m, n, *alpha, *A, lda, *B, ldb, *C, ldc)	S/D/C/Z	Triangular matrix-matrix multiplication (C = alpha A*B)
cublas?trsm(handle, side, ul, op, di, m, n, *alpha, *A, lda, *B, ldb)	S/D/C/Z	Solves the system of equations given in the triangular matrix

The arguments of these functions closely resemble those of the Level 2 functions. The `op`, `opA`, and `opB` arguments identify the operation to be performed on a matrix before it's used. `ul` specifies if the matrix is upper or lower triangular, and `di` identifies if the main diagonal contains unit values (ones).

Every function in the table accepts matrix dimensions, and it's important to see what they signify:

- `m` — the number of rows in **A** and **C**
- `n` — the number of columns in **B** and **C**
- `k` — the number of columns of **A** and rows of **B**

The `cublas?gemm` function performs basic matrix-matrix multiplication and addition. If `opA` and `opB` are set to `CUBLAS_OP_N`, the operation's equation will be given by $\mathbf{C} = \alpha\mathbf{AB} + \beta\mathbf{C}$. Matrix-matrix multiplication isn't commutative, so the order of **A**, **B**, and **C** is important.

The `cublas?symm` and `cublas?hemm` functions have a `side` argument that configures the order of the multiplication of the **A** and **B** matrices. This can be set to one of two values:

- `CUBLAS_SIDE_LEFT` — the equation is $\mathbf{C} = \alpha\mathbf{AB} + \beta\mathbf{C}$
- `CUBLAS_SIDE_RIGHT` — the equation is $\mathbf{C} = \alpha\mathbf{BA} + \beta\mathbf{C}$

The rank update functions in this table are similar to those in the preceding table, but instead of performing rank-1 and rank-2 updates, they perform rank-k and rank-2k updates. For `cublas?syrk`, the equation for the rank-k update is $\mathbf{C} = \alpha\mathbf{AA}^T + \beta\mathbf{C}$. For `cublas?syr2k`, the equation for a rank-2k update is $\mathbf{C} = \alpha\mathbf{AB}^T + \alpha\mathbf{BA}^T + \beta\mathbf{C}$.

The last two functions involve triangular matrices. The `cublas?trmm` function performs matrix-matrix multiplication, but unlike `cublas?gemm` and `cublas?hemm`, there's no addition step. It simply stores the multiplication result into the **C** matrix.

6.5 Summary

Linear algebra is a critical part of many computational tasks, including graphical rendering, wireless communication, and machine learning. Researchers created a standard set of linear algebra operations called BLAS, and cuBLAS is Nvidia's implementation of these operations. Once you get used to the data types and naming convention, cuBLAS is very easy to work with.

The simplest BLAS operations are the Level 1 BLAS operations, which involve vectors. To perform these operations, cuBLAS provides functions for dot products, norms, and rotations. The example project shows how to perform complex dot products using the `cublasCdotc` and `cublasCdotu` functions.

Level 2 BLAS operations involve a matrix and a vector. Most of the functions perform matrix-vector multiplication and rank updates. The example project demonstrates how the `cublasSgemv` function can be used to compute $\alpha Ax + \beta y$, where A is a matrix, x and y are vectors, and α and β are constants.

Level 3 BLAS operations involve two matrices, and most of the functions perform matrix-matrix multiplication, rank updates, and system solution. This chapter hasn't provided any example code for Level 3 operations, but the next chapter will show how systems of equations can be solved using the cuSOLVER and cuDSS libraries.

Chapter 7

Solving Systems of Equations with cuSOLVER and cuDSS

Throughout science, finance, and engineering, many problems require solving systems of linear equations. These equations can be expressed in matrix form, and the goal is to obtain the values of the x vector in the equation Ax = b, where A is a matrix of coefficients.

The nature of A depends on the interaction of the variables. If every variable interacts with every other variable, A will contain few zero elements. This type of matrix is called a *dense matrix*. However, if each variable only interacts with a few variables, A will contain a large percentage of zeros. If this percentage is large enough, A is called a *sparse matrix*.

Nvidia's primary library for solving a system of equations represented by a dense matrix is cuSOLVER. The functions of this library are straightforward to use, and closely resemble those of the cuBLAS library discussed in Chapter 6. This chapter explains how these functions operate and presents an example project that demonstrates how they're used.

For sparse matrix systems, this chapter presents a new package called cuDSS (Direct Sparse Solver). This always reaches an exact solution for a well-conditioned sparse matrix system, and can scale across multiple GPUs. With cuDSS, there's no need for preconditioning and equations can be solved for multiple right-hand side vectors. That is, once Ax = b has been solved for x, it's very easy to solve $Ax_1 = b_1$ for x_1 and $Ax_2 = b_2$ for x_2.

Nvidia strongly recommends using cuDSS for sparse matrices, but it does have disadvantages. cuDSS functions require a lot of memory, and if a sparse matrix is too large, direct solution may not be possible. Also, cuDSS isn't provided in the CUDA Toolkit, and must be downloaded separately. This chapter explains how to obtain the cuDSS library and demonstrates how its functions can be used.

7.1 Introducing cuSOLVER

Working with cuSOLVER is like working with the cuBLAS package described in Chapter 6. If your application targets a single-GPU system, the build process requires two steps:

1. Include the appropriate header file in your source code.
2. Link the application to the cuSOLVER library (libcusolver.so on Linux, cusolver.lib on Windows) in the build procedure.

The header file depends on the type of matrices you're dealing with and your execution platform. If you look through the include folder in the CUDA Toolkit, you'll find several header files provided by cuSOLVER, including cusolverDn.h, cusolverSp.h, cusolverRf.h, and cusolverMg.h.

At the time of this writing, most cuSOLVER functions are deprecated. The multi-GPU functions in cusolverMg.h are replaced with the functions in cusolverMp.h, which is provided in Nvidia's HPC SDK (not discussed in this book). The sparse matrix functions in cusolverSp.h and cusolverRf.h are replaced with the functions in cudss.h, which will be discussed later in this chapter.

The first part of this chapter is concerned with the functions and data structures in cusolverDn.h, which haven't been deprecated. The "Dn" stands for dense, and the functions are used to solve systems of equations that are represented by dense matrices. In contrast, the functions in cuDSS are used to solve systems of equations represented by sparse matrices.

Many developers get confused about the difference between dense and sparse matrices, and expect there to be a simple dividing line. But there's no clear distinction. If a matrix contains so many zeros that it's faster to process nonzero elements, use sparse-matrix functions. If it doesn't contain that many zeros, use dense-matrix functions.

7.1.1 General Development

At a high level, the process of working with cuSOLVER consists of five steps:

1. Call `cusolverDnCreate` to initialize a `cusolverDnHandle_t`.
2. Create data structures to store input and output values.
3. Create a workspace to hold data used during computation.
4. Call one or more cuSOLVER functions to process matrices.
5. Deallocate the handle by calling `cusolverDnDestroy`.

cuSOLVER functions return a `cusolverStatus_t` value that identifies the operation's status. Table 7.1 lists each of the possible status values and provides a description of each.

Table 7.1
Status Values

Name	Description
CUSOLVER_STATUS_SUCCESS	The operation completed successfully
CUSOLVER_STATUS_NOT_INITIALIZED	The cuSOLVER library was not initialized properly
CUSOLVER_STATUS_ALLOC_FAILED	Failed to allocate resources
CUSOLVER_STATUS_INVALID_VALUE	A value passed to a function was invalid or unsupported
CUSOLVER_STATUS_ARCH_MISMATCH	Attempted to access capability unsupported on the target device
CUBLAS_STATUS_EXECUTION_FAILED	The program failed to execute
CUBLAS_STATUS_INTERNAL_ERROR	An internal operation failed, usually related to asynchronous operation
CUBLAS_STATUS_MATRIX_TYPE_ NOT_SUPPORTED	An invalid matrix descriptor was provided to the function

Unfortunately, cuSOLVER doesn't provide a function that accepts a status code and returns a string description of the error. For this reason, the example code in this chapter contains a custom function named `checkcuSolverErr`. Its signature is given as follows:

```
inline void checkcuSolverErr(cusolverStatus_t status,
  const char* file, int line)
```

This executes a `switch` statement that checks the status code. For each error code (greater than 0), the function prints a message that identifies the nature of the error, the name of the file, and the line on which the error occurred.

This function isn't called directly, but is invoked through the `checkcuSolverError` macro, which is defined in the following way:

```
#define checkcuSolverError(err) (checkcuSolverErr((err),
  __FILE__, __LINE__))
```

7.1.2 Naming Conventions

The names of the cuSOLVER functions discussed in this chapter all have the same basic structure:

```
cusolverDn<type><op_name>
```

cuSOLVER's functions only accept matrices that contain floating-point values. But these values may be real or complex and they may be single-precision or double-precision. Chapter 6 provided a brief explanation of real and complex arithmetic.

These functions don't use generics to support different types of arguments. Instead, `cusolverDn` is followed by a capital letter that identifies the type and precision of its data:

- `S` — real-valued single-precision
- `D` — real-valued double-precision
- `C` — complex-valued single-precision
- `Z` — complex-valued double-precision

After the type, a function's name is determined by the nature of the operation it performs. In general, these names are taken from LAPACK (Linear Algebra Package), which is a popular open-source package for solving linear algebra systems.

The name of a LAPACK operation has four, five, or six characters given in the format `pmma`, `pmmaa`, or `pmmaaa`. The `p` identifies the type of the matrix elements (`S`, `D`, `C`, `Z`) and the `mm` identifies the matrix type (`ge` for general, `sy` for symmetric, `he` for Hermitian, and so on). The last part of the name identifies the nature of the operation to be performed. Three important operations are given as follows:

- `trf` — LU or Cholesky factorization (depending on the matrix type)
- `qrf` — QR factorization
- `trs` — solve a system using factorization

An example will help to make this clear. The function that factorizes a general dense double-precision matrix using LU decomposition is `cusolverDnDgetrf`. The `Dn` indicates that the target matrix is dense and the following `D` indicates the matrix contains real-valued double-precision values. The `ge` indicates that the matrix is general (not symmetric) and the `trf` indicates that the function performs LU factorization.

Similarly, the function that solves a dense matrix system after factorization is `cusolverDnDgetrs`. The example code in this section will use both of these functions to solve a system of equations.

7.2 Dense Matrix Operations

A matrix is considered dense if the number of zero elements isn't significant enough to warrant using sparse matrix operations. In general, if less than 80% of a matrix's elements are zero, the matrix is considered dense.

In the CUDA Toolkit, cusolverDn.h declares the functions that operate on dense matrices. These functions perform a wide range of operations, including solving systems of equations, matrix inversion, and eigenvalue/eigenvector problems. This chapter focuses on solving systems of linear equations like the one illustrated in Figure 7.1:

$$
\begin{aligned}
a_{11}x_1 + a_{12}x_2 + a_{13}x_3 + a_{14}x_4 &= b_1 \\
a_{21}x_1 + a_{22}x_2 + a_{23}x_3 + a_{24}x_4 &= b_2 \\
a_{31}x_1 + a_{32}x_2 + a_{33}x_3 + a_{34}x_4 &= b_3 \\
a_{41}x_1 + a_{42}x_2 + a_{43}x_3 + a_{44}x_4 &= b_4
\end{aligned}
$$

Figure 7.1 A Linear System of Equations

These four equations all involve the same set of unknowns: x_1, x_2, x_3, and x_4, and the goal of the solver is to compute their values. The general process involves treating the constants a_{ij} as elements of a matrix A, the variables x_i as elements of a vector x, and the constants b_j as the elements of a vector b.

In this manner, the system of equations can be expressed with the equation Ax = b. Figure 7.2 shows what this looks like graphically.

$$
\begin{vmatrix} a_{11} & a_{12} & a_{13} & a_{14} \\ a_{21} & a_{22} & a_{23} & a_{24} \\ a_{31} & a_{32} & a_{33} & a_{34} \\ a_{41} & a_{42} & a_{43} & a_{44} \end{vmatrix}
\begin{vmatrix} x_1 \\ x_2 \\ x_3 \\ x_4 \end{vmatrix}
=
\begin{vmatrix} b_1 \\ b_2 \\ b_3 \\ b_4 \end{vmatrix}
$$

Figure 7.2 Equations in Matrix Form

Finding the inverse of A can take a great deal of time and effort, so solvers factorize (or decompose) A into matrices that are easier to work with. To be more specific, solvers decompose rectangular matrices into triangular matrices, and then solve the system using substitution.

The cuSOLVER functions rely on three main methods for matrix decomposition. You don't need to know the low-level details of how these algorithms work, but when choosing a function, it's important to know when one method is preferred over another:

- **LU decomposition** — factorizes into triangular matrices L and U, good choice for general dense matrices
- **Cholesky decomposition** — accelerates LU decomposition for symmetric positive definite (SPD) matrices
- **QR decomposition** — factorizes into triangular matrices Q and R, good choice for rectangular (non-square) matrices and ill-conditioned matrices

A matrix is symmetric if it's square and it equals its transpose. That is, each element equals the element reflected across the diagonal ($a_{ij} = a_{ji}$). A matrix is positive definite if pre-multiplying and post-multiplying it by a non-zero vector produces a positive number. If a matrix is symmetric and positive definite (SPD), an application should use Cholesky decomposition to solve the system.

If a matrix is rectangular but not square, QR decomposition is the preferred choice. This is slower but more stable than LU decomposition, so if a matrix is ill-conditioned (sensitive to small changes in values), QR decomposition is the algorithm to use.

LU decomposition is a good choice for most systems. This section explains how this algorithm works and then shows how it can be performed in code using cuSOLVER.

7.2.1 The LU Decomposition Algorithm

LU decomposition is one of the most important algorithms in linear algebra, and it's simple to understand. It iterates through each column of the matrix (left to right) and operates on the submatrix bounded by the column and the row with the same index. This operation consists of four tasks:

1. Find the index of the value with the greatest magnitude in the column and set this value as the *pivot*.
2. Swap the row containing the pivot with the row with the same index as the current column.
3. Scale the elements in the current column by 1/pivot.
4. Find the product of the current row and matrix and subtract it from the submatrix.

To understand how this works, it helps to walk through an example. This discussion shows how LU decomposition converts the following 4-by-4 matrix into a lower-triangular matrix (L) and an upper-triangular matrix (U):

$$\begin{vmatrix} 9 & 4 & 2 \\ 2 & 6 & 5 \\ 4 & 1 & 7 \end{vmatrix}$$

In the first column, the largest value (9) is chosen as the pivot. This is already in the first row, so there's no need to swap rows. Dividing the other elements in the column by 9, $m_{21} = 2/9$ and $m_{31} = 4/9$. The next step multiplies Row 1 by m_{21} and subtracts the result from Row 2, and multiplies Row 1 by m_{31} and subtracts the result from Row 3. Replacing the rows in the matrix with the subtracted results produces the following matrix:

$$\begin{vmatrix} 9 & 4 & 2 \\ 0 & 46/9 & 41/9 \\ 0 & -7/9 & 55/9 \end{vmatrix}$$

In Column 2, the value with the largest magnitude is 46/9. This is already in the second row so there's no need to swap. Dividing the last value in the column by the pivot, $m_{32} = -7/46$. The next step multiplies Row 2 by m_{32} and subtracts the result from Row 3. Replacing Row 3 with the result produces the following:

$$\begin{vmatrix} 9 & 4 & 2 \\ 0 & 46/9 & 41/9 \\ 0 & 0 & 313/46 \end{vmatrix}$$

This matrix is upper triangular because all the values below the diagonal equal 0. Now that we've constructed the U matrix, we can construct the lower-triangular matrix L by setting the diagonal values to 1 and inserting m_{21}, m_{31}, and m_{32} below the diagonal. This produces the following result.

$$\begin{vmatrix} 1 & 0 & 0 \\ 2/9 & 1 & 0 \\ 4/9 & -7/46 & 1 \end{vmatrix}$$

If you compute the product of L and U, you'll see that it equals the original matrix. Once these values are obtained, Ax = b can be transformed into the equation (LU)x = b, which is straightforward to solve using simple algebraic substitution.

7.2.2 Solving Dense Systems with cuSOLVER

To solve a system of equations like the one shown above with cuSOLVER, it's important to be familiar with three functions:

- `cusolverDnDgetrf_bufferSize` — determines how much memory to allocate for the solver
- `cusolverDnDgetrf` — performs LU decomposition
- `cusolverDnDgetrs` — uses the result of LU decomposition to solve the system of matrices

In CUDA, a *workspace* is a portion of device memory set aside to store intermediate data during computation. If an application needs to perform similar operations repeatedly, it only needs to create the workspace once and then free the device memory when the operations are complete.

Before an application can create a workspace, it needs to know how much memory to allocate. For LU decomposition, this information is provided by the `cusolverDnDgetrf_bufferSize` function, whose signature is given as follows:

```
cusolverDnDgetrf_bufferSize(cusolverDnHandle_t handle,
  int num_rows, int num_cols, double *A, int lda, int *work_sz);
```

The arguments are straightforward to understand. The third parameter, `A`, points to the block of device memory that holds a matrix whose dimensions are given by `num_rows` and `num_cols`. `lda` stands for leading dimension of A, and identifies the number of elements between consecutive columns of a matrix. If there's no padding, this can be set to `num_rows`. The last parameter identifies the number of elements of the workspace needed to process the given matrix.

Once the workspace memory has been allocated, the application can perform LU decomposition by calling `cusolverDnDgetrf`. Its signature is given as follows:

```
cusolverDnDgetrf(cusolverDnHandle_t handle, int num_rows,
  int num_cols, double *A, int lda, double *workspace_mem,
  int *pivot_mem, int *err_mem)
```

The last two parameters deserve explanation. As discussed earlier, LU decomposition requires swapping rows of the A matrix, which is called *pivoting*. The `pivot_info` parameter points to device memory that stores which rows have been swapped. A solution function requires this in order to restore the rows to their original order.

Some matrices can't be inverted, which means the system of equations can't be solved. These are called *singular matrices*, and LU decomposition of a singular matrix will fail. The last parameter, `err_mem`, will contain 0 if decomposition was successful. If the value is greater than 0, the matrix is singular and the value identifies which element on the diagonal of the U matrix equals 0.

If LU decomposition is successful, the application can use the results to solve the system of equations. This is made possible by the `cusolverDnDgetrs` function, whose signature is given as follows:

```
cusolverDnDgetrs(cusolverDnHandle_t handle,
  cublasOperation_t trans, int dim, int nrhs, const double *A,
  int lda, const int *pivot_mem, double *B, int ldb, int *err_mem)
```

The second argument identifies what operation should be applied to the A matrix before the processing starts. This can be set to one of three values:

- `CUBLAS_OP_N` — No operation should be applied
- `CUBLAS_OP_T` — A will be set to its transpose, A^T
- `CUBLAS_OP_C` — A will be set to its conjugate transpose, A^H

This function can be used to solve multiple systems that have the same A matrix. That is, it can be used to solve $Ax_1 = b_1$ for x_1, $Ax_2 = b_2$ for x_2, and $Ax_3 = b_3$ for x_3. To make this possible, the b vectors must be combined (as columns) of the B matrix.

The fourth argument, `nrhs`, identifies the number of columns of the B matrix. If only one solution is required, this can be set to 1.

The `pivot_mem` argument stores the pivot information generated during LU decomposition, and `err_mem` stores information about the matrix singularity. The `B` parameter points to the block of device memory containing the B matrix, and `ldb` is the number of elements between adjacent columns of the B matrix.

When the solution is computed, the result will be stored in the B matrix. In other words, when Ax = b is solved for x, the values of x can be retrieved by accessing the block of memory that contained b.

7.2.3 Example Application

The ch07_dense_solver project demonstrates how a system of equations can be solved using the three functions discussed earlier. Figure 7.3 illustrates the A matrix, x vector, and b vector corresponding to the equations to be solved.

$$\begin{vmatrix} 9 & 4 & 2 \\ 2 & 6 & 5 \\ 4 & 1 & 7 \end{vmatrix} \begin{vmatrix} x_0 \\ x_1 \\ x_2 \end{vmatrix} = \begin{vmatrix} 3 \\ 5 \\ 7 \end{vmatrix}$$

Figure 7.3 Example System of Equations

The project's code is in main.cu, and it performs nine important operations:

1. Calls `cusolverDnCreate` to initialize a `cusolverDnHandle_t`.
2. Allocates blocks of device memory for the A matrix and b vector.
3. Initializes the A matrix and b vector on the host and copies the data to the device.
4. Determines the required size of the workspace by calling `cusolverDnDgetrf_bufferSize`.
5. Allocates memory for the workspace, pivot data, and error data.
6. Performs LU decomposition by calling `cusolverDnDgetrf`.
7. Solves the system of equations by calling `cusolverDnDgetrs`.
8. Copies the values of the b vector to the host and displays the results.
9. Frees device memory and destroys the handle with `cusolverDnDestroy`.

Listing 7.1 presents the code of the main.cu file in the ch07_dense_solver project. Keep in mind that the matrix data is provided in column-major format.

Listing 7.1 ch07_dense_solver/main.cu

```
int main() {

  // Create cuSOLVER handle
  cusolverDnHandle_t handle;
  checkcuSolverError(cusolverDnCreate(&handle));

  // Allocate device memory
  const int dim = 3;
  double *aMem, *bMem;
  checkCudaErrors(cudaMalloc(reinterpret_cast<void**>(&aMem),
    dim * dim * sizeof(double)));
  checkCudaErrors(cudaMalloc(reinterpret_cast<void**>(&bMem),
    dim * sizeof(double)));

  // Define the A matrix data
  double aData[dim * dim] = {
    9.0, 2.0, 4.0, // Column 0
    4.0, 6.0, 1.0, // Column 1
    2.0, 5.0, 7.0  // Column 2
  };

  // Set the b vector data
  double bData[dim] = { 3.0, 5.0, 7.0 };

  // Transfer data to device
  checkCudaErrors(cudaMemcpy(aMem, aData,
    dim * dim * sizeof(double), cudaMemcpyHostToDevice));
  checkCudaErrors(cudaMemcpy(bMem, bData,
    dim * sizeof(double), cudaMemcpyHostToDevice));

  // Determine workspace size
  int workspace_size = 0;
  checkcuSolverError(cusolverDnDgetrf_bufferSize(handle,
    dim, dim, aMem, dim, &workspace_size));

  // Allocate memory for workspace
  double *workspaceMem;
  checkCudaErrors(cudaMalloc(reinterpret_cast<void**>(
    &workspaceMem), workspace_size * sizeof(double)));
```

Listing 7.1 ch07_dense_solver/main.cu (Continued)

```
    // Allocate memory for pivot info and error info
    int *pivotMem, *errMem;
    checkCudaErrors(cudaMalloc(reinterpret_cast<void**>(&pivotMem),
      dim * sizeof(int)));
    checkCudaErrors(cudaMalloc(reinterpret_cast<void**>(&errMem),
      sizeof(int)));

    // Perform LU decomposition
    checkcuSolverError(cusolverDnDgetrf(handle, dim, dim, aMem,
      dim, workspaceMem, pivotMem, errMem));

    // Solve matrix system
    checkcuSolverError(cusolverDnDgetrs(handle, CUBLAS_OP_N, dim,
      1, aMem, dim, pivotMem, bMem, dim, errMem));

    // Copy solution to host
    double xData[dim];
    checkCudaErrors(cudaMemcpy(xData, bMem, dim * sizeof(double),
      cudaMemcpyDeviceToHost));

    // Display solution
    std::cout << "Solution: x = " << xData[0] << ", " << xData[1]
      << ", " << xData[2] << std::endl;

    // Free resources
    cudaFree(aMem);
    cudaFree(bMem);
    cudaFree(workspaceMem);
    cudaFree(pivotMem);
    cudaFree(errMem);

    // Destroy the cuSOLVER handle
    cusolverDnDestroy(handle);
    return 0;
}
```

Most of this code should be straightforward to understand. In addition to allocating memory for input data, it creates memory blocks for the workspace, pivot information, and matrix singularity result. The size of the workspace is determined by calling `cusolverDnDgetrf_bufferSize` with the dimensions of the matrix.

For the pivot information, the LU decomposition produces an array of integers that identify which rows were swapped. In the example listing, the block of device memory is allocated with the following code:

```
cudaMalloc(reinterpret_cast<void**>(&pivotMem),
  dim * sizeof(int));
```

As shown, the application allocates enough memory to store one integer for each row of the matrix. For the matrix singularity data, the application only needs to allocate a single integer.

After performing LU decomposition with `cusolverDnDgetrf`, the application calls `cusolverDnDgetrs` to solve the system of equations. This accepts the A matrix, the b vector, the pivot information, and the matrix singularity information. The second argument is set to `CUBLAS_OP_N`, so the A matrix won't be transformed before processing starts.

After the equations are solved, the results are stored in the block of memory containing the b vector. After calling `cudaMemcpy` to transfer this to the host, the application displays the results, which should equal 0.118211, 0.0191693, and 0.929712.

7.3 Sparse Matrix Operations

When scientists and engineers model systems based on differential equations, the result is a huge matrix with a high proportion of zeros. These matrices are called *sparse matrices*, and because there are so many zeros, it's more efficient to use operations that process the non-zero elements individually.

An example will clarify the relationship between differential equations and sparse matrices. In structural engineering, one of the most important differential equations is the Euler-Bernoulli equation, which relates the deflection of a beam, w(x) to its supported load, q. This is given as follows:

$$\frac{d^2}{dx^2}\left(EI\frac{d^2w}{dx^2}\right) = q$$

This is easy to solve if you're modeling a single beam. But if you have a complex system of interconnected elements, it becomes difficult to arrive at an exact solution. For example, if you had to model the tower illustrated in Figure 7.4, how would you determine the deflections of its beams?

Figure 7.4 Assembly of Beam Elements

To model real-world systems like this tower, scientists and engineers rely on finite element analysis, or FEA. The basic procedure consists of five steps:

1. Divide the system into simple elements that are easy to analyze.
2. Obtain the differential equation or equations for each element.
3. Convert the differential equations into linear equations using an approximation method such as the Petrov-Galerkin method or the Ritz-Galerkin method.
4. Combine the linear equations into a matrix.
5. Solve the system of equations to obtain an approximate solution.

The third step is particularly important. FEA approximates differential equations, which are hard to solve, using linear equations that are easy to solve. A system like a transmission tower may have hundreds or even thousands of these equations, but each element only connects to two or three others. Therefore, each equation will only have two or three nonzero terms.

When these equations are combined, the resulting system can be solved using matrix operations. Figure 7.5 depicts a matrix that corresponds to a structural model of a real transmission tower.

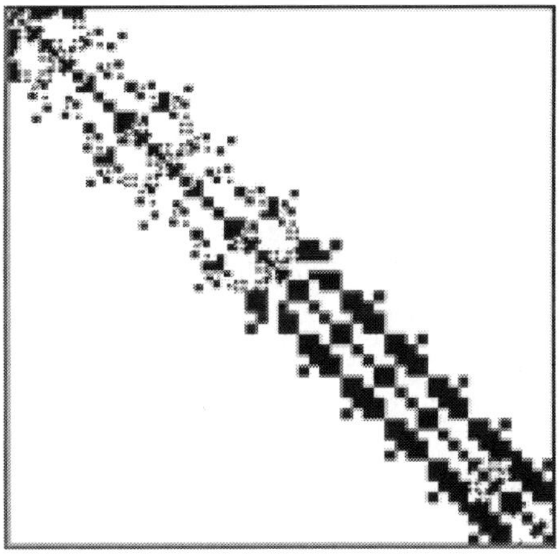

Figure 7.5 Elements of a Sparse Matrix

The matrix size is 153-by-153, but only 10% (2,423) of its elements are nonzero. These are represented by dots, and the matrix is symmetric because the model is symmetric.

There's no clear distinction between sparse matrices and dense matrices. A matrix is considered sparse if its proportion of zero-to-nonzero elements, called its *sparsity*, makes it worth our while to use processing routines specifically created for sparse matrices. This discussion presents the functions in cuSOLVER specifically suited for solving sparse matrix systems.

7.3.1 Direct and Iterative Solvers

Nvidia's packages for sparse matrices can be confusing. cuSPARSE supports simple operations like multiplying sparse matrices or converting data from one format to another. But it doesn't provide functions that can solve a system of equations represented by a sparse matrix. That is, cuSPARSE can multiply a sparse matrix A by a vector x, but it can't solve Ax = b for x.

In general, there are two main ways to solve a system of equations represented by a sparse matrix:

- **direct solver** — factors the matrix using methods similar to LU decomposition and QR factorization discussed earlier, usually requires a great deal of memory
- **iterative solver** — starts with a guess and repeatedly reduces the error, not guaranteed to reach a solution, requires preconditioning

Choosing between these methods is an important decision. Direct solvers will always reach a solution if the matrix isn't singular, but the decomposition produces matrices that aren't sparse. The insertion of nonzero values is called *fill-in*, and if a sparse matrix is very large, this can require a great deal of memory.

Because direct solvers require so much memory, FEA applications traditionally rely on iterative solvers. Instead of factorizing the sparse matrix, these solvers start with an initial estimate for the x vector, compute the Ax product, and compare the result to b. Based on the comparison, these solvers continue updating the estimate until the error reaches a minimum.

Iterative solvers don't need a lot of memory, and computing the Ax product can be performed quickly on parallel systems like GPUs. But these solvers aren't guaranteed to reach a correct answer, and they have major issues with ill-conditioned matrices.

As GPU memory increases, direct solvers become more viable for solving sparse matrix systems. This explains why Nvidia has spent so much effort developing cuDSS (CUDA Direct Sparse Solver). If you want to use an iterative solver, you can call the functions in cusolverSp.h, which work like the cuSOLVER functions discussed earlier. However, this discussion focuses on cuDSS development.

7.3.2 The CUDA Direct Sparse Solver (cuDSS)

The cuDSS package changes so rapidly that Nvidia doesn't include it in the CUDA Toolkit. It can be downloaded from developer.nvidia.com/cudss-downloads. It provides a header file (cudss.h) and libraries (cudss.lib and cudss64_0.dll on Windows, libcudss.so on Linux). On Windows, I recommend copying cudss.h to the toolkit's include folder, cudss.lib to the toolkit's lib/x64 folder, and cudss64_0.dll to the toolkit's bin folder. On Linux, it's a good idea to check ch07_sparse_solver/CMakeLists.txt to make sure the cuDSS library is located where the compiler will find it.

The functions in cudss.h aren't like the functions in any package discussed so far. Solving a sparse matrix system (Ax = b) with cuDSS requires at least five steps:

1. Create the initial data structures (`cudssHandle_t`, `cudssConfig_t`, and `cudssData_t`).
2. Create matrix structures for A, x, and b by calling functions like `cudssMatrixCreateDn` and `cudssMatrixCreateCsr`.
3. Call `cudssExecute` multiple times to compute fill-in, decompose the sparse matrix, and solve the matrix system.
4. Copy the solution results in the x matrix to the host.
5. Free resources and destroy the `cudssHandle_t`, `cudssConfig_t`, and `cudssData_t`.

Initial Data Structures

Every cuDSS application starts by calling three functions that create data structures required for cuDSS operations. The first is `cudssCreate`, which initializes a `cudssHandle_t` that holds the cuDSS library context. This must be called before any other cuDSS functions.

```
cudssHandle_t handle;
cudssCreate(&handle);
```

Next, an application needs to create an object that holds settings for the solver. These settings are contained in a `cudssConfig_t` structure, and it's initialized with `cudssConfigCreate`.

```
cudssConfig_t config;
cudssConfigCreate(&config);
```

Once the `cudssConfig_t` structure is initialized, an application can configure the solver's operation by calling `cudssConfigSet` with a parameter name, a value for the parameter, and the number of bytes in the value. The function's signature is given as follows:

```
cudssConfigSet(cudssConfig_t config, cudssConfigParam_t param,
  void *value, size_t sizeInBytes);
```

These configuration parameters tell the solver how to operate. Table 7.2 lists the configuration parameters available and provides a description of each.

Table 7.2

Configuration Parameter Names

Parameter Name	Description
CUDSS_CONFIG_REORDERING_ALG	Algorithm to be employed for reordering
CUDSS_CONFIG_FACTORIZATION_ALG	Algorithm to be employed for factorization
CUDSS_CONFIG_SOLVE_ALG	Algorithm to be employed for solving
CUDSS_CONFIG_USE_MATCHING	Whether to perform matching during preprocessing (disabled by default)
CUDSS_CONFIG_MATCHING_ALG	Algorithm to be employed for matching (if enabled)

CUDSS_CONFIG_SOLVE_MODE	The transformation to be applied to the system matrix (not supported)
CUDSS_CONFIG_IR_N_STEPS	Number of steps during iterative refinement
CUDSS_CONFIG_IR_TOL	Iterative refinement tolerance (not supported)
CUDSS_CONFIG_PIVOT_TYPE	Type of pivoting to be employed
CUDSS_CONFIG_PIVOT_THRESHOLD	Threshold that determines when diagonal elements are subject to pivoting
CUDSS_CONFIG_PIVOT_EPSILON	Value used to replace small diagonal elements that arise during factorization
CUDSS_CONFIG_MAX_LU_NNZ	Upper limit on the number of nonzero entries in LU factors
CUDSS_CONFIG_HYBRID_MODE	Memory mode (device-only or host/device)
CUDSS_CONFIG_HYBRID_DEVICE_MEMORY_LIMIT	Maximum memory to be used in hybrid mode
CUDSS_CONFIG_USE_CUDA_REGISTER_MEMORY	Whether to use mapped memory in hybrid mode
CUDSS_CONFIG_HOST_NTHREADS	Number of threads used in multithreaded mode
CUDSS_CONFIG_HYBRID_EXECUTE_MODE	Whether to use hybrid mode for execution
CUDSS_CONFIG_PIVOT_EPSILON_ALG	Algorithm to compute the pivot epsilon
CUDSS_CONFIG_ND_NLEVELS	Minimum number of levels for the nested dissection reordering
CUDSS_CONFIG_UBATCH_SIZE	Number of matrices in a uniform batch of systems
CUDSS_CONFIG_UBATCH_INDEX	Index of the matrix in a uniform batch
CUDSS_CONFIG_USE_SUPERPANELS	Whether to use superpanel optimization
CUDSS_CONFIG_DEVICE_COUNT	Number of devices available
CUDSS_CONFIG_DEVICE_INDICES	Indices of available devices
CUDSS_CONFIG_SCHUR_MODE	Enables/disables Schur complement mode
CUDSS_CONFIG_DETERMINISTIC_MODE	Enables/disables deterministic mode

Because cuDSS is so new, some of these options are either unsupported or only partially supported. It's interesting to note that, unlike cuSOLVER, cuDSS doesn't allow applications to select the factorization algorithm, such as LU decomposition or Cholesky decomposition. Instead, the application identifies the properties of the matrix, and cuDSS chooses the best algorithm by itself.

The `CUDSS_CONFIG_FACTORIZATION_ALG` provides a limited amount of control. However, this can only be set to one of two values: `CUDSS_ALG_DEFAULT` or `CUDSS_ALG_1`. Nvidia's documentation doesn't explain how `CUDSS_ALG_1` works, so it's a good idea to use the default algorithm.

cuDSS supports using a matching algorithm to reorder rows and columns so that large values are moved to the diagonal. This can improve the matrix's stability and the accuracy of the solution, but it slows down the computation and requires extra memory. Matching is disabled by default, but can be enabled with the following code:

```
// Create the configuration structure
cudssConfig_t config;
cudssConfigCreate(&config);

// Set the parameter value to 1
int enable = 1;
cudssConfigSet(config, CUDSS_CONFIG_MATCHING_TYPE, &enable,
  sizeof(int));
```

Another interesting feature is hybrid mode, which allows cuDSS to access memory on the host as well as memory on the GPU. If the `CUDSS_CONFIG_HYBRID_MODE` parameter is assigned to 1, cuDSS will store portions of the factorization results in host memory. The amount of memory that can be accessed is determined by the `CUDSS_CONFIG_HYBRID_DEVICE_MEMORY_LIMIT` parameter.

Once the `cudssConfig_t` is created and configured, the application can initialize a `cudssData_t` structure. This holds intermediate data generated during computation, such as matrix factors, and pointers to user data. Initialization is accomplished with the `cudssDataCreate` function, and the following code shows how it can be used:

```
cudssData_t solver_data;
cudssDataCreate(handle, &solver_data);
```

Unlike configuration options, applications usually shouldn't change the content of the `cudssData_t` structure. Instead, they can access information about the solver's data by calling `cudssDataGet`. Its signature is given as follows:

```
cudssDataGet(cudssHandle_t handle, cudssData_t data,
  cudssDataParam_t param, void *value, size_t size,
  size_t *sizeWritten)
```

This accepts a parameter name and a pointer to the variable to contain the parameter's value. The parameter can be set to any of the values listed in Table 7.3.

Table 7.3

Data Property Names

Property Name	Data Provided
CUDSS_DATA_INFO	Error information
CUDSS_DATA_LU_NNZ	Number of nonzero elements in LU factors
CUDSS_DATA_NPIVOTS	Number of pivots required during factorization
CUDSS_DATA_INERTIA	Indices of inertia for symmetric/Hermitian non positive-definite matrices
CUDSS_DATA_PERM_REORDER_ROW	Row permutation after reordering
CUDSS_DATA_PERM_REORDER_COL	Column permutation after reordering
CUDSS_DATA_PERM_ROW	Final row permutation
CUDSS_DATA_PERM_COL	Final column permutation
CUDSS_DATA_DIAG	Diagonal of the factorized matrix
CUDSS_DATA_USER_PERM	User permutation to be applied
CUDSS_DATA_HYBRID_DEVICE_MEMORY_MIN	Minimum amount of device memory required for hybrid mode
CUDSS_DATA_COMM	Communicator to be used for multi-GPU multi-node (MGMN) mode
CUDSS_DATA_MEMORY_ESTIMATES	Memory estimates for host and device memory required for the current memory mode
CUDSS_DATA_PERM_MATCHING	Permutation used during matching
CUDSS_DATA_SCALE_ROW	Row scaling of the factorized matrix
CUDSS_DATA_SCALE_COL	Column scaling of the factorized matrix
CUDSS_DATA_NSUPERPANELS	Number of superpanels in the matrix
CUDSS_DATA_USER_SCHUR_INDICES	Schur complement indices provided by user
CUDSS_DATA_SCHUR_SHAPE	Shape of the Schur complement matrix
CUDSS_DATA_SCHUR_MATRIX	Schur complement matrix
CUDSS_DATA_USER_ELIMINATION_TREE	Information about the elimination tree to be used instead of reordering
CUDSS_DATA_ELIMINATION_TREE	Separator sizes that make up the elimination tree
CUDSS_DATA_USER_HOST_INTERRUPT	Pointer to a user interrupt that can be used to interrupt cuDSS routines

By accessing these parameters with cudssDataGet, applications can obtain information about the matrix factorization that can be later used to tune the factorization process. For example, the following function returns the number of pivots that were required during factorization:

```
int numPivots = 0;
size_t sizeWritten = 0;

// Access cuDSS data
cudssDataGet(handle, data, CUDSS_DATA_NPIVOTS, &numPivots,
  sizeof(int), &sizeWritten);
```

For most applications, the cudssData_t structure is read-only. But some parameters can be used to provide data to the solver. For example, calling cudssDataSet with CUDSS_DATA_USER_PERM provides a user-defined permutation. Users can also provide information related to Schur complement matrices and elimination trees.

Creating Dense and Sparse Matrices

After initializing the basic data structures for cuDSS, an application needs to create matrix structures for the system. If the goal is to solve Ax = b for x, the application will need to create matrix structures for A, x, and b. To be precise, the goal is to create a cudssMatrix_t structure from data stored in blocks of device memory.

In the system Ax = b, x and b can be thought of as dense matrices with a single column. In cuDSS, these matrices can be created by calling cudssMatrixCreateDn, whose signature is given as follows:

```
cudssMatrixCreateDn(cudssMatrix_t *matrix, int64_t nrows,
  int64_t ncols, int64_t ld, void *data,
  cudaDataType_t valueType, cudssLayout_t layout)
```

Most of these arguments are easy to understand. The second and third arguments set the matrix dimensions. The fourth, ld, sets the number of bytes between the start of consecutive columns for column-major matrices. The fifth argument, data, points to the region of device memory containing the matrix elements.

The sixth parameter, valueType, identifies the data type of the matrix elements. This can be set to values like CUDA_R_16F or CUDA_R_32F for real values, CUDA_C_16F or CUDA_C_32F for complex values, or CUDA_R_8I or CUDA_R_32I for integer values. The last parameter of the function must be set to CUDSS_LAYOUT_COL_MAJOR because cuDSS doesn't support row-major matrices.

Creating a `cudssMatrix_t` for a sparse matrix is significantly more involved. The function to know is `cudssMatrixCreateCsr`, and its signature is given as follows:

```
cudssMatrixCreateCsr(cudssMatrix_t *matrix, int64_t nrows,
  int64_t ncols, int64_t nnz, void *rowStart, void *rowEnd,
  void *colIndices, void *values, cudaDataType_t indexType,
  cudaDataType_t valueType, cudssMatrixType_t mtype,
  cudssMatrixViewType_t mview, cudssIndexBase_t indexBase)
```

In the function name, `Csr` stands for Compressed Sparse Row, which is a popular format for storing elements of a sparse matrix. This format is only concerned with the nonzero elements of the matrix, whose number must be given with the `nnz` argument. These values must be provided in the device memory block referenced by the `values` argument, and must be provided row-by-row (not column-by-column).

The `colIndices` argument must be set to an array (stored on the device) that contains the column index of each non-zero value. Therefore, if the matrix has `nnz` nonzero values, `colIndices` will reference a memory block containing `nnz` integers.

The `rowStart` and `rowEnd` arrays are more complex. The goal is to identify which values belong to which row of the matrix. The ith element of `rowStart` identifies the first element index that belongs to row i. The ith element of `rowEnd` identifies the last element index that belongs to row i.

For example, if the ith element of `rowStart` is N, then the Nth value in `values` is the *first* value in Row i. If the ith element of `rowStart` is M, then the Mth value in `values` is the *last* value in Row i.

To understand the CSR format, it helps to look at a simple example. The 4-by-4 matrix illustrated in Figure 7.6 has seven nonzero elements, so `nnz` should be set to `7`. These elements are [5.0, 2.0, 1.0, 3.0, 2.0, 9.0, 7.0], so this is the array corresponding to the function's `values` argument.

$$
\begin{vmatrix}
5.0 & 0.0 & 2.0 & 0.0 \\
1.0 & 3.0 & 0.0 & 0.0 \\
0.0 & 2.0 & 9.0 & 0.0 \\
0.0 & 0.0 & 0.0 & 7.0
\end{vmatrix}
$$

Figure 7.6 An Example Sparse Matrix

The `colIndices` array stores the column index of each nonzero element. In this case, the array will equal [0.0, 2.0, 0.0, 1.0, 1.0, 2.0, 3.0].

Row 0 contains the first and second nonzero elements, so `rowStart[0]` = `0` and `rowEnd[0]` = `1`. Row 1 contains the third and fourth nonzero elements, so `rowStart[1]` = `2` and `rowEnd[1]` = `3`. Row 2 contains the fifth and sixth nonzero elements, so `rowStart[2]` = `4` and `rowEnd[2]` = `5`. Row 3 contains the seventh nonzero element, so `rowStart[3]` = `6` and `rowEnd[3]` = `6`.

The `indexType` argument identifies the data type of the index values, and for 32-bit integers, this should be set to `CUDA_R_32I`. The `valueType` argument identifies the data type of the matrix values, and for real 32-bit floating-point values, this should be set to `CUDA_R_32F`.

The `mtype` argument identifies the type of the overall matrix, and can be set to one of five values:

1. `CUDSS_MTYPE_GENERAL` — general matrix, the solver will use LDU factorization with optional local or global pivoting
2. `CUDSS_MTYPE_SYMMETRIC` — real symmetric matrix, the solver will use LDL^T factorization with optional local pivoting
3. `CUDSS_MTYPE_SPD` — symmetric positive-definite matrix, the solver will use Cholesky factorization with optional local pivoting
4. `CUDSS_MTYPE_HERMITIAN` — complex Hermitian matrix, the solver will use LDL^H factorization with optional local pivoting
5. `CUDSS_MTYPE_HPD` — Hermitian positive-definite matrix, the solver will use complex Cholesky factorization with optional local pivoting

The `mview` argument identifies which portion of the matrix should be processed. This can be set to one of three values:

- `CUDSS_MVIEW_FULL` — the entire matrix should be processed
- `CUDSS_MVIEW_LOWER` — the lower triangle of the matrix (including the diagonal) should be processed
- `CUDSS_MVIEW_UPPER` — the upper triangle of the matrix (including the diagonal) should be processed

The last argument of `cudssMatrixCreateCsr` is `indexBase`, which identifies whether row and column indices start at 0 (following C/C++ convention) or 1 (following FORTRAN convention). This argument can be set to `CUDSS_BASE_ZERO` or `CUDSS_BASE_ONE`. In this chapter's example code, this argument will always be set to `CUDSS_BASE_ZERO`.

Solver Execution

cuSOLVER provides multiple functions for matrix factorization and system solution, but cuDSS provides only one: `cudssExecute`. Its signature is given as follows:

```
cudssExecute(cudssHandle_t handle, int phase,
  cudssConfig_t solverConfig, cudssData_t solverData,
  cudssMatrix_t inputMatrix, cudssMatrix_t solution,
  cudssMatrix_t rhs)
```

Most of the argument types should look familiar. The `cudssHandle_t` is the overall handle for cuDSS, the `cudssConfig_t` stores configuration settings, and `cudssData_t` stores data properties.

The second argument, `phase`, identifies the operation to be performed. cuDSS supports several different phases, but for most applications, only three are necessary:

1. `CUDSS_PHASE_ANALYSIS` — performs reordering and symbolic factorization
2. `CUDSS_PHASE_FACTORIZATION` — decomposes matrix into factors using an algorithm determined by the matrix type
3. `CUDSS_PHASE_SOLVE` — solves the system of equations

The `cudssExecute` function needs to be called once for each phase. If the A matrix is stored in `aMem`, the x vector is stored in `xMem`, and the b vector is stored in `bMem`, the following code will execute the analysis phase:

```
cudssExecute(handle, CUDSS_PHASE_ANALYSIS, config, data, aMem,
  xMem, bMem)
```

The analysis phase performs two important tasks. First, it comes up with a permutation vector that will reduce the number of nonzero elements that need to be added during factorization (*fill-in*). This is called *reordering*. Then, using the permutation vector, it determines the layout of the matrix's factors. This is called *symbolic factorization*.

Once the analysis phase is completed, an application can call `cudssDataGet` with the `cudssData_t`. This provides information about the permutation vector and properties of the matrix's factors.

After calling `cudssExecute` for the factorization and solve phases, an application can access the solution by copying the appropriate matrix to the host. In the `cudssExecute` function, this corresponds to the `solution` argument. The code in the following discussion will demonstrate how this can be accomplished.

7.3.3 Example Application

The code in the ch07_sparse_solver project demonstrates how cuDSS can be used to solve a sparse matrix system. The application starts by reading the content of a sparse matrix file obtained from the Harwell-Boeing collection. Therefore, before I present the code, I'd like to explain what these files are and how they're formatted.

The Harwell-Boeing Collection

Most real-world sparse matrices are huge, with hundreds and thousands of rows. You can see what I mean by visiting the Harwell-Boeing collection at http://math.nist.gov/MatrixMarket/data/Harwell-Boeing. Each link identifies a set of files whose sparse matrices were generated through system analysis, from structural engineering to nuclear reactor analysis to economics and demographics.

This chapter focuses on the MCCA matrix, which is a real-valued, unsymmetric 180-by-180 matrix with 2,659 nonzero entries. The mcca.mtx file is provided as part of the MatrixMarket repository sponsored by the National Institute for Standards and Technology (NIST). These resources are freely available for public use, and I'd like to thank everyone at NIST and Boeing for making them available.

The file format consists of three sections:

- **banner** — the matrix format and numeric properties (real or complex, symmetric or general)
- **dimensionality** — the number of rows, number of columns, and number of nonzero elements
- **data** — the row, column, and value of each nonzero element

For example, the first seven lines of mcca.mtx are as follows:

```
%%MatrixMarket matrix coordinate real general
180 180 2659
1 1  1.0580564100000e+02
2 1  2.3250942200000e+05
3 1  2.6991455000000e+06
4 1  1.3705616800000e+02
1 2  9.1345272100000e+00
```

As given in the second line, this file lists the 2,659 nonzero elements of a 180-by-180 matrix. Each line following the second line provides the row index, column index, and value of a nonzero element in the matrix. It's important to note that the row index and column index start with 1, not 0. These values need to be decremented in code.

Another important point is that the matrix elements aren't sorted by row. This is a significant concern because cuDSS expects the matrix data to be sorted. In the example code, each entry is represented by a tuple and the collection of entries is declared with the following vector:

```
std::vector<std::tuple<int, int, double>> entries(nnz);
```

After opening the file, the example code reads the entries into this vector, sorts them by row, and then forms the arguments needed to call cuDSS routines. The following discussion explains how this is accomplished.

The ch07_sparse_solver Project

The code in ch07_sparse_solver demonstrates how cuDSS functions can be used to solve a sparse matrix system. The goal is to compute the x vector in the equation Ax = b, where the elements of A are given in mcca.mtx. The elements of b are computed so that every element in x is 1.0.

Inside the project, the code in main.cu performs eleven operations:

1. Opens the mcca.mtx file. Skips the first line and any lines starting with `%`.
2. Reads the number of rows, number of columns, and number of nonzero elements in the sparse matrix.
3. Reads each entry and writes the values to the `entries` vector.
4. Sorts the `entries` vector and constructs the data structures needed for cuDSS.
5. Allocates blocks of device memory for the cuDSS arguments, including the A matrix, x vector, b vector, and row/column information.
6. Creates the three fundamental cuDSS data structures (`cudssHandle_t`, `cudssConfig_t`, and `cudssData_t`).
7. Calls `cudssMatrixCreateCsr` to create a sparse `cudssMatrix_t` for the A matrix.
8. Calls `cudssMatrixCreateDn` to create dense `cudssMatrix_t` structures for the x vector and b vector.
9. Calls `cudssExecute` three times. Once to analyze the system, once to factorize the A matrix, and once to solve the system of equations.
10. Copies the x vector to the host and checks that each element equals 1.0.
11. Destroys the cuDSS data structures and frees the blocks of device memory.

Listing 7.2 presents the code of the main function in the main.cu source file. It uses two functions for error checking. `checkCudaErrors` receives the status of CUDA operations and `checkcudssError` examines the status of cuDSS operations.

Listing 7.2 ch07_sparse_solver/main.cu

```
int main() {
  // Load MTX file
  std::ifstream infile("mcca.mtx");
  if (!infile) {
    std::cerr << "Error: could not open matrix file"
      << std::endl;
    return 1;
  }

  // Skip comments
  std::string line;
  while (std::getline(infile, line)) {
    if (line.empty() || line[0] == '%') continue;
    break;
  }

  // Read matrix dimensions
  std::stringstream ss(line);
  int rows, cols, nnz;
  ss >> rows >> cols >> nnz;

  // Create data containers on host
  std::vector<int> rowStarts(rows + 1, 0);
  std::vector<int> colIndices(nnz);
  std::vector<double> values(nnz);
  std::vector<double> rhs(rows, 0.0);
  std::vector<std::tuple<int, int, double>> entries(nnz);

  // Create vector entries
  int rowIndex, colIndex;
  double value;
  for (int i = 0; i < nnz; i++) {
    infile >> rowIndex >> colIndex >> value;
    entries[i] = std::make_tuple(rowIndex - 1,
      colIndex - 1, value);
    rowStarts[rowIndex]++;
    rhs[rowIndex - 1] += value;
  }
  infile.close();
```

Listing 7.2 ch07_sparse_solver/main.cu (Continued)

```
// Update rowStarts array
for (int i = 1; i < rowStarts.size(); ++i) {
  rowStarts[i] += rowStarts[i - 1];
}

// Sort file entries
std::sort(entries.begin(), entries.end(),
  [](const auto& a, const auto& b) {
    return std::get<0>(a) < std::get<0>(b);
  });

// Get column indices and values
for (int i = 0; i < nnz; i++) {
  colIndices[i] = std::get<1>(entries[i]);
  values[i] = std::get<2>(entries[i]);
}

// Allocate device memory
int *rowMem, *colMem;
double *valMem, *xMem, *bMem;
checkCudaErrors(cudaMalloc((void**)&rowMem,
  (rows + 1) * sizeof(int)));
checkCudaErrors(cudaMalloc((void**)&colMem,
  nnz * sizeof(int)));
checkCudaErrors(cudaMalloc((void**)&valMem,
  nnz * sizeof(double)));
checkCudaErrors(cudaMalloc((void**)&xMem,
  cols * sizeof(double)));
checkCudaErrors(cudaMalloc((void**)&bMem,
  rows * sizeof(double)));

// Copy data to device
checkCudaErrors(cudaMemcpy(rowMem, rowStarts.data(),
  (rows + 1) * sizeof(int), cudaMemcpyHostToDevice));
checkCudaErrors(cudaMemcpy(colMem, colIndices.data(),
  nnz * sizeof(int), cudaMemcpyHostToDevice));
checkCudaErrors(cudaMemcpy(valMem, values.data(),
  nnz * sizeof(double), cudaMemcpyHostToDevice));
checkCudaErrors(cudaMemcpy(bMem, rhs.data(),
  rows * sizeof(double), cudaMemcpyHostToDevice));
```

Listing 7.2 ch07_sparse_solver/main.cu (Continued)

```
// Declare cuDSS data structures
cudssHandle_t handle;
cudssConfig_t config;
cudssData_t solver_data;

// Initialize data structures
checkcudssError(cudssCreate(&handle));
checkcudssError(cudssConfigCreate(&config));
checkcudssError(cudssDataCreate(handle, &solver_data));

// Create matrices
cudssMatrix_t aMat, xMat, bMat;
checkcudssError(cudssMatrixCreateCsr(&aMat, rows, cols,
  nnz, rowMem, NULL, colMem, valMem, CUDA_R_32I, CUDA_R_64F,
  CUDSS_MTYPE_GENERAL, CUDSS_MVIEW_FULL, CUDSS_BASE_ZERO));

// Create Vector x
checkcudssError(cudssMatrixCreateDn(&xMat, rows, 1, rows, xMem,
  CUDA_R_64F, CUDSS_LAYOUT_COL_MAJOR));

// Create Vector b
checkcudssError(cudssMatrixCreateDn(&bMat, rows, 1, rows, bMem,
  CUDA_R_64F, CUDSS_LAYOUT_COL_MAJOR));

// Perform reordering and symbolic factorization
checkcudssError(cudssExecute(handle, CUDSS_PHASE_ANALYSIS,
  config, solver_data, aMat, xMat, bMat));

// Perform factorization
checkcudssError(cudssExecute(handle, CUDSS_PHASE_FACTORIZATION,
  config, solver_data, aMat, xMat, bMat));

// Solve the system
checkcudssError(cudssExecute(handle, CUDSS_PHASE_SOLVE, config,
  solver_data, aMat, xMat, bMat));

// Copy results to host
std::vector<double> xData(rows);
checkCudaErrors(cudaMemcpy(xData.data(), xMem,
  rows * sizeof(double), cudaMemcpyDeviceToHost));
```

Listing 7.2 ch07_sparse_solver/main.cu (Continued)

```
  // Check results
  bool pass = true;
  for (int i = 0; i < rows; i++) {
    if (std::abs(xData[i] - 1.0) > 0.01) {
      pass = false;
      break;
    }
  }
  std::cout << (pass ? "Solution check passed" :
    "Solution check failed") << std::endl;

  // Destroy data structures
  cudssMatrixDestroy(aMat);
  cudssMatrixDestroy(xMat);
  cudssMatrixDestroy(bMat);
  cudssDataDestroy(handle, solver_data);
  cudssConfigDestroy(config);
  cudssDestroy(handle);

  // Free device memory
  cudaFree(rowMem);
  cudaFree(colMem);
  cudaFree(valMem);
  cudaFree(xMem);
  cudaFree(bMem);
  return 0;
}
```

It's particularly important to understand how the sparse matrix structure is created. This is accomplished with the following code:

```
cudssMatrixCreateCsr(&aMat, rows, cols, nnz, rowMem, NULL,
  colMem, valMem, CUDA_R_32I, CUDA_R_64F, CUDSS_MTYPE_GENERAL,
  CUDSS_MVIEW_FULL, CUDSS_BASE_ZERO)
```

The `rows`, `cols`, and `nnz` arguments are taken from the second line of the MTX file. The locations and values of the nonzero elements are given in the blocks of memory containing `rowMem`, `colMem`, and `valMem`. The following arguments identify the indices as 32-bit integers (`CUDA_R_32I`) and the values as 64-bit double-precision values (`CUDA_R_64F`). The matrix isn't symmetric, so the matrix type is set to `CUDSS_MTYPE_GENERAL`. The solver should process the entire matrix, so the next argument is set to `CUDSS_MVIEW_FULL`.

After creating a structure for the sparse matrix, the application calls `cudssMatrixCreateDn` to create dense matrices for the x vector and the b vector. The x vector is initially empty.

The values in the b vector are computed such that every value in the x vector is 1.0. This is easy to do, since b[i] is the sum of the elements in the ith row of the A matrix. For example, the first value of b, b[0], equals the sum of the elements of the first row of the A matrix.

After creating matrix structures, the application calls `cudssExecute` three times. The first call performs reordering and symbolic factorization. The second call performs the actual factorization, and the third call uses the factors to solve the system of equations. When the solution is finished, the application reads the values of the x vector, compares each to 1.0, and displays the test result.

7.4 Summary

Throughout science and engineering, many problems require solving for the vector x in the equation Ax = b. When A is large, solving the system can take a great deal of time. For this reason, CUDA provides specialized libraries tailored to the structure of the matrix. If A is dense, cuSOLVER provides functions for solving the system. If A is sparse, Nvidia recommends using the functions in cuDSS (Direct Sparse Solver).

The first part of this chapter discussed cuSOLVER in detail, including its data structures and naming convention. This part also walked through the process of LU factorization, which decomposes A into a lower-triangular matrix and an upper-triangular matrix. The example project uses cuSOLVER to solve a very simple system of equations represented as a 3-by-3 matrix.

The second part of this chapter focuses on sparse matrices and explains why they appear so frequently in engineering. The sparse matrix functions in cuSOLVER have been deprecated, so this discussion has explored the routines in the new cuDSS library. These routines require matrix elements to be provided in CSR (Compressed Sparse Row) format, and the example project shows how a sparse matrix can be converted from MatrixMarket format to CSR and then factorized with cuDSS.

Chapter 8

Frequency Analysis with cuFFT

Throughout science and engineering, there's no escape from the Fast Fourier Transform (FFT). Cell phones, audio players, X-ray scanners, radar receivers, and biometric scanning systems all depend on this algorithm to extract frequency information from signals that vary over time and space. The nature of the signal may change, but one constant remains true: processing speed is a major priority.

To perform the FFT on Nvidia GPUs, the CUDA Toolkit provides the cuFFT library. The goal of this chapter is to present the theory underlying frequency analysis and show how it can be implemented using cuFFT functions.

In my opinion, the simplest way to understand the FFT is to study the Discrete Fourier Transform (DFT). The first part of this chapter presents the DFT and shows how the operation can be inverted to obtain the inverse Discrete Fourier Transform, or IDFT.

The second part of the chapter explores the cuFFT library and presents the most important functions needed for frequency analysis. Unlike other libraries discussed so far, cuFFT requires that applications create a *plan* that defines the nature of the transform. To be specific, the plan requires the size of the FFT, the data type of its elements, and the batch size.

After an application has created a plan, it can call cuFFT functions to execute the FFT and IFFT. cuFFT provides several execution functions for different types of input and output. The last part of this chapter shows how to use these functions to perform the forward FFT on real-valued input data and then perform the inverse FFT on the complex results to obtain the original input.

8.1 The Discrete Fourier Transform (DFT)

Before I delve into the details of using cuFFT, it's important to have a basic understanding of Fourier analysis. This section presents the theory and terminology underlying the Discrete Fourier Transform (DFT), which is much simpler to understand than the FFT.

In 1822, the French physicist Joseph Fourier published a revolutionary paper entitled *The Analytic Theory of Heat*. This presented a method of decomposing complex mathematical relationships into a sum of simple functions. Over the centuries, his method has been applied to a variety of applications, and though Fourier focused on continuous functions, our goal is to analyze the content of digital values called *samples*.

Before we get into the math, it's important to establish the notation that will be used throughout this chapter:

- **N** — number of samples in the input signal
- **x(n)** — array of samples where n is the sample index
- **Δt** — time between samples, measured in seconds
- **T** — duration of the input signal, measured in seconds
- X(k) — the frequency components of the signal, where k identifies the frequency index
- **Δf** — the frequency interval between the components in X(k), measured in Hertz

Don't be concerned if you don't fully grasp the theory presented in this section. But you should have a basic familiarity with the subject so that you can write code that analyzes a signal's frequency using cuFFT.

8.1.1 Dot Products and the DFT

The fundamental operation of the DFT is the *dot product*, which is used throughout engineering and computer science. As discussed in Chapter 6, a dot product takes two arrays of equal length, multiplies their corresponding elements together, and computes the sum of the products. This plays a central role in this discussion, so on the following page, Figure 8.1 illustrates two arrays named x(i) and w(i).

Multiplying the values in the two arrays produces sixteen values: 0.2 × 0.0, 0.1 × 0.4, 0.5 × 0.7, 0.6 × 0.9, 0.7 × 1.0, 0.3 × 0.9, 0.0 × 0.7, −0.1 × 0.4, −0.2 × 0.0, −0.1 × −0.4, −0.5 × −0.7, −0.4 × −0.9, −0.7 × −1.0, 0.0 × −0.9, 0.3 × −0.7, and 0.1 × −0.4. Adding these products together gives us the dot product, which equals 3.06.

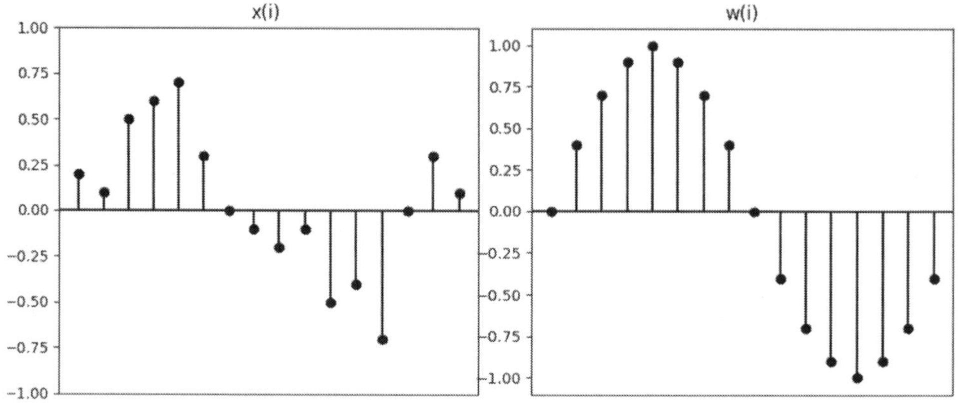

Figure 8.1 Arrays in a Dot Product (First Example)

Figure 8.2 illustrates arrays named y(i) and w(i). In this case, the sixteen products add up to a dot product of -0.23. The values of w(i) haven't changed, but if you compare the graphs of x(i) and y(i), it's clear that x(i) resembles w(i) more than y(i) does.

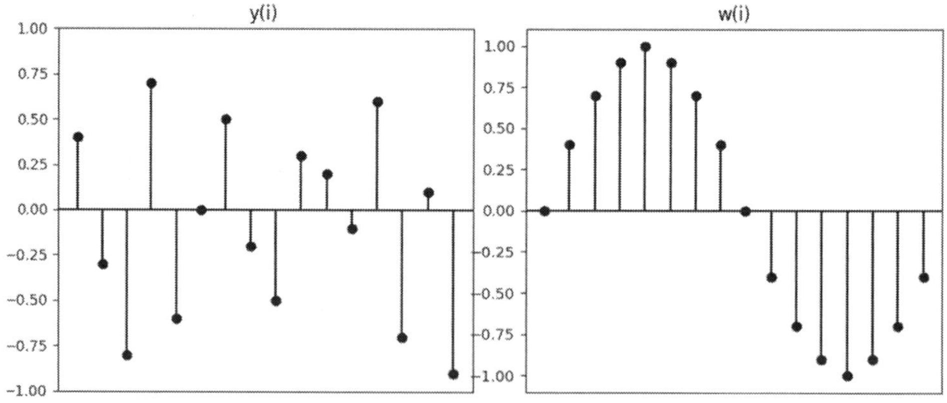

Figure 8.2 Arrays in a Dot Product (Second Example)

This leads to a general observation: *a larger dot product implies that the two arrays are similar, while a smaller dot product implies that the arrays are different.* In this case, x(i) resembles w(i) more than y(i) resembles w(i), so the dot product of x(i) and w(i) is larger than the dot product of y(i) and w(i).

As you may recall from trigonometry, the values of w(i) have a specific shape called a *sine wave*. We'll be dealing a lot with sine waves and cosine waves in this book, so for convenience, we'll refer to these waves in general as *sinusoids*.

The main point to know about sinusoids is that *every sinusoid has a single frequency*. Its frequency depends on how often the sinusoid reaches successive peaks (or *oscillates*) in a given time. If a sinusoid oscillates once per second, it has a frequency of 1 Hertz (abbreviated Hz). Figure 8.3 illustrates three sine waves: 1 Hz, 2 Hz, and 3 Hz.

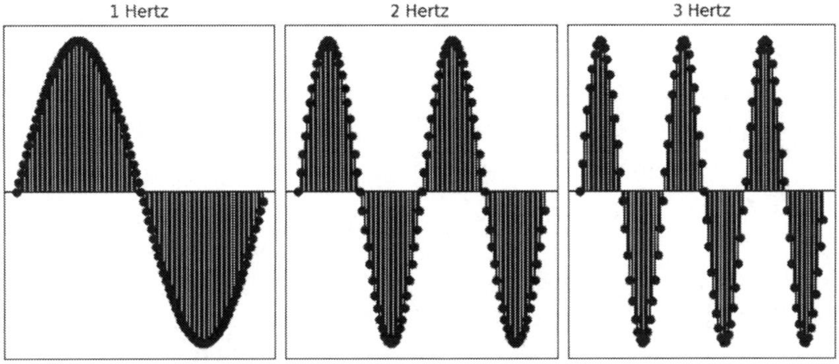

Figure 8.3 Sine Waves at Different Frequencies

When we take the dot product of an audio signal and a sinusoid of frequency f, the result tells us how much the audio signal resembles the sinusoid. This is significant because this dot product is the *frequency component at frequency f*.

The goal of the DFT is to find dot products of a signal with sinusoids at every frequency of interest. But we can't compute dot products for every possible frequency. Therefore, we need to answer an important question: which frequencies should we be interested in?

8.1.2 Frequencies of Interest

The goal of the DFT is to compute a finite number of frequency components. It would be nice if we could pick an arbitrary set of frequencies, but the DFT algorithm makes the selection for us. The values of the DFT's frequencies depend on the period of the audio signal (T) and the number of audio samples (N).

The first frequency of interest, called the *fundamental frequency* or f_1, has a frequency equal to 1/T. This represents a single oscillation over the course of the signal, so if a signal only oscillates once during its period, f_1 will be significant.

Further frequencies of interest, f_k, are obtained by multiplying the fundamental frequency by k. This is given mathematically by the following equation:

$$f_k = kf_1 = \frac{k}{T}$$

Just as f_k identifies the frequency corresponding to k oscillations in a period, f_0 identifies the frequency corresponding to no oscillations at all. This represents a frequency of 0 Hz. If a signal's value is constant over the course of its period, then it doesn't oscillate at all. In this case, X(0) is significant and components at higher frequencies will equal 0.

For general signals, the highest frequency of interest is f_{N-1}, where N is the number of samples in x(n). This gives us N frequencies of interest: f_0 through f_{N-1}. If the samples all have real values, it's only necessary to compute components of frequencies up to $f_{N/2}$. This is called the *Nyquist frequency*.

8.1.3 Complex Sinusoids

Chapter 6 explained that a complex number is the sum of a real number and an imaginary number. Similarly, the definition of a sinusoid can be expanded such that a sinusoid is any sum of a real-valued cosine wave and an imaginary-valued sine wave, where the cosine and sine have the same frequency. The general equation for a sinusoid with frequency f is:

$$w_f(t) = \cos(2\pi ft) - i\sin(2\pi ft)$$

Figure 8.4 illustrates a sinusoid at the fundamental frequency, f_1. At this frequency, a sinusoid only oscillates once during the length of the signal, T.

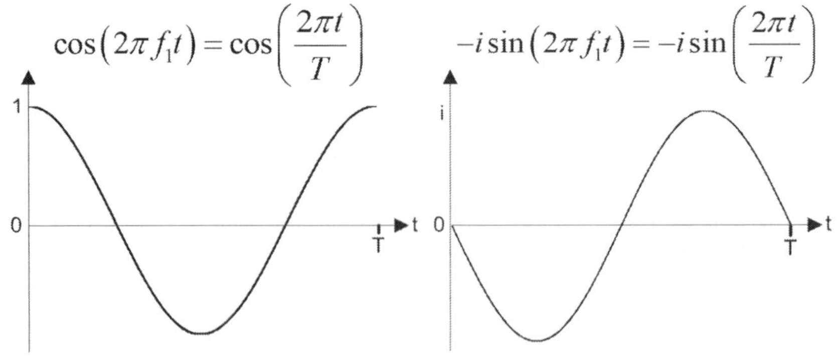

Figure 8.4 Sinusoid at the Fundamental Frequency (Continuous)

In this figure, the arguments of the cosine and sine are $2\pi f_1 t$, where t is time and f_1 is the frequency. To make sense of this, remember that the cosine and sine accept values in radians, and both oscillate through one cycle as the radian measure runs from 0 to 2π. Therefore, if the argument is 2πt, the sinusoid represents the frequency that oscillates once as time increases from 0 to 2π.

The DFT requires a finite number of sinusoids and each sinusoid must have a finite number of values. For this reason, we need to rework the expression for w so that it depends only on n and k. We know that our frequencies of interest, $f_k = k/T$, and we know that $t = n\Delta t$. Replacing this in the sinusoid equation produces the following:

$$w_f(t) = \cos\left(\frac{2\pi k(n\Delta t)}{T}\right) - i\sin\left(\frac{2\pi k(n\Delta t)}{T}\right)$$

We can relate the duration of the signal, T, to the total number of samples, N, by noting that $T = N\Delta t$. Replacing this in the equation, we arrive at a final expression:

$$w_k(n) = \cos\left(\frac{2\pi nk}{N}\right) - i\sin\left(\frac{2\pi nk}{N}\right)$$

This gives us a vector with N values, and Figure 8.5 depicts $w_k(n)$ from n = 0 to N–1. This discrete sinusoid corresponds to the fundamental frequency, given by k = 1.

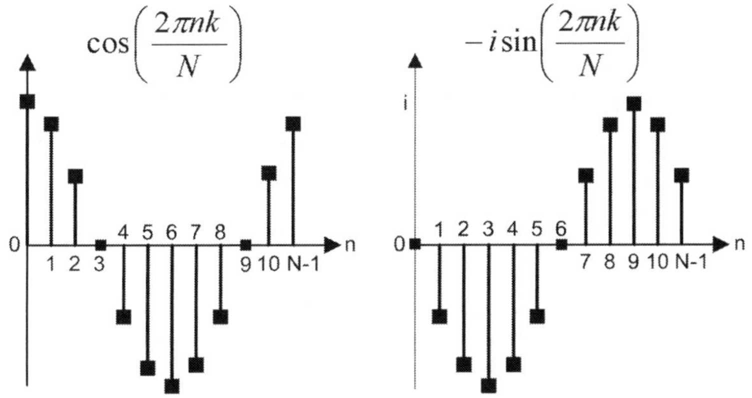

Figure 8.5 Sinusoid at the Fundamental Frequency (Discrete)

In this graph, neither the cosine nor the sine returns to its starting value. Instead, they proceed from 0 to N–1.

8.1.4 The DFT and IDFT Equations

The DFT computes X(k) by taking the dot product of x(n) and w_k(n), where k runs from 0 to N–1. This can be expressed with the DFT equation:

$$X(k) = x(n) \cdot w_k(n) = \sum_{n=0}^{N-1} x(n)\left[\cos\left(\frac{2\pi nk}{N}\right) - i\sin\left(\frac{2\pi nk}{N}\right)\right]$$

The relationship between X(k) and x(n) can also be expressed using matrix-vector multiplication. If w_k(n) form the rows of a matrix W as k runs from 0 to N–1, the matrix-vector product, Wx = X, can be expressed in the following way:

$$\begin{vmatrix} w_0[0] & w_0[1] & w_0[2] & \cdots & w_0[N-1] \\ w_1[0] & w_1[1] & w_1[2] & \cdots & w_1[N-1] \\ w_2[0] & w_2[1] & w_2[2] & \cdots & w_2[N-1] \\ \vdots & \vdots & \vdots & & \vdots \\ w_{N-1}[0] & w_{N-1}[1] & w_{N-1}[2] & \cdots & w_{N-1}[N-1] \end{vmatrix} \begin{vmatrix} x[0] \\ x[1] \\ x[2] \\ \vdots \\ x[N-1] \end{vmatrix} = \begin{vmatrix} X[0] \\ X[1] \\ X[2] \\ \vdots \\ X[N-1] \end{vmatrix}$$

The W matrix is called the *DFT matrix*. After scaling, W is unitary and symmetric, which means that its inverse, W^{-1}, equals its conjugate, W*. The conjugate of a matrix can be obtained by reversing the sign of each imaginary term. For the DFT, finding the inverse means converting the negative imaginary sine in w_k(n) into a positive imaginary sine.

The inverse matrix is critical because it can be used to compute the *inverse DFT*, or IDFT. The IDFT transforms frequency components to time-domain samples, and can be expressed in matrix-vector form as x = W^{-1}X. The IDFT equation is given as follows:

$$x(n) = X(k) \cdot w_k^*(n) = \frac{1}{N}\sum_{k=0}^{N-1} X(k)\left[\cos\left(\frac{2\pi nk}{N}\right) + i\sin\left(\frac{2\pi nk}{N}\right)\right]$$

The w_k^*(n) term refers to the conjugate of w_k(n), whose negative sine values, –i sin(x), are replaced with positive sine values, i sin(x). The only other difference between this and the DFT equation is the scaling factor 1/N. This is necessary to ensure that, once x(n) is transformed into X(k), the inverse transformation will produce the original x(n).

A common alternative is to scale the DFT equation and IDFT equations by 1/√N. This ensures that the two equations are symmetric and that the matrix represented by the DFT equation is unitary.

8.2 Using the cuFFT Package

Now that you understand the goal of cuFFT, you're ready to access the library in code. If your application targets a single GPU, the development and build process requires two steps:

- In the source code, include the cufft.h header file.
- In the build file, link the cuFFT library (libcufft.so on Linux, cufft.lib on Windows) into the build procedure.

Coding an application with cuFFT is similar to coding applications with cuBLAS and cuSOLVER. There are five main steps involved:

1. Create data structures to contain input and output data.
2. Declare a `cufftHandle` structure.
3. Create a plan that identifies the nature of the transform to be performed.
4. Use the plan to execute the transform.
5. Call `cufftDestroy` to destroy the `cufftHandle`.

This section discusses each of these steps and then walks through the development of an example that uses cuFFT to perform the FFT and IFFT.

8.2.1 Creating a Plan

The cuFFT library doesn't use a single algorithm to perform FFTs. Instead, it chooses the algorithm according to a plan that must be defined by the application. This plan needs to identify several aspects about the desired operation, including the following:

- Transform dimensions (e.g., length of the array).
- Floating-point precision (single or double).
- The type of input data (real or complex)
- The type of output data (real or complex)
- Batch size: how many FFTs to run in parallel
- Memory layout

The cuFFT library provides four functions that define FFT plans. Table 8.1 lists their signatures and provides a description of each.

Table 8.1

Plan Creation Functions

Function	Description
`cufftPlan1d(cufftHandle *plan, int nx, cufftType type, int batch)`	Create a plan to transform one-dimensional data
`cufftPlan2d(cufftHandle *plan, int nx, int ny, cufftType type)`	Create a plan to transform two-dimensional data
`cufftPlan3d(cufftHandle *plan, int nx, int ny, int nz, cufftType type)`	Create a plan to transform three-dimensional data
`cufftPlanMany(cufftHandle *plan, int rank, int *n, int *inembed, int istride, int idist, int *onembed, int ostride, int odist, cufftType type, int batch)`	Create a plan that executes multiple transforms on structured data in one, two, or three dimensions

There is no data structure that specifically represents an FFT plan. Instead, these functions configure the `cufftHandle` with the information needed to execute the desired transform. A `cufftHandle` is just an integer, and when you create a plan, you set it to a value that tells cuFFT which algorithm to execute.

After the `cufftHandle`, each function accepts the dimensions of the data to be transformed. This is given as `nx` for one-dimensional data, `nx` and `ny` for two-dimensional data, and `nx`, `ny`, and `nz` for three-dimensional data.

The transform's type is given by the `type` argument, which can be set to one of six values:

- `CUFFT_R2C` — single-precision real to single-precision complex
- `CUFFT_C2R` — single-precision complex to single-precision real
- `CUFFT_C2C` — single-precision complex to single-precision complex
- `CUFFT_D2Z` — double-precision real to double-precision complex
- `CUFFT_Z2D` — double-precision complex to double-precision real
- `CUFFT_Z2Z` — double-precision complex to double-precision complex

cuFFT makes it possible to execute multiple transforms on data in the same block of device memory. This is configured by setting the `batch` parameter to identify the number of transforms to be performed.

For example, if the `batch` argument of `cufftPlan1d` is set to 10, cuFFT will perform ten transforms on the provided data. The block of device memory is expected to contain ten arrays, each containing `nx` values. The first operation transforms the first `nx` values, the second transforms the second `nx` values, and so on.

The last function in the table, `cufftPlanMany`, provides more options that identify how multiple transforms should be executed. The second argument, `rank`, identifies the number of dimensions in the data (1, 2, or 3) and the second argument, `n`, points to the array that identifies the number of values in each dimension.

If the arrays contain padding, then the physical size of each array will exceed the number of values times the number of bytes in each value. In this case, the physical sizes should be provided in the `inembed` and `onembed` arguments. The number of elements between successive values to be transformed should be given in the `istride` and `ostride` arguments. The `idist` and `odist` arguments specify the number of elements between the end of one transform and the start of the next.

To clarify how these functions are used, it helps to look at an example. Suppose your application needs to perform a simple 512-point FFT on single-precision real-valued data. You could create the plan with the following code:

```
cufftPlan1d(&handle, 512, CUFFT_R2C, 1)
```

Now suppose the block of device memory contains six arrays of 512 values each, and the application should transform each array separately. This can be easily accomplished by setting the `batch` value to `6`.

```
cufftPlan1d(&handle, 512, CUFFT_R2C, 6)
```

If each input value is followed by a value to be ignored, then the stride between successive values is 2. In this case, the application needs to call `cufftPlanMany` with the following code:

```
int n[1] = {512};
cufftPlanMany(&handle, 1, n, NULL, 2, 1024, NULL, 1, 257,
  CUFFT_R2C, 6);
```

If each input value in each array is followed by an ignored value, then the number of values in each transform is $2 \times 512 = 1024$. This is given in the `idist` argument.

Because the input is real-valued, the output doesn't need to store all 512 complex values for each array. Due to symmetry, it only needs to store $N/2 + 1$ values. This is why the `odist` argument is set to `257`.

8.2.2 Executing the Transform

Once the plan is created, cuFFT makes it straightforward to transform data in a block of device memory. In fact, applications can use a plan to perform multiple transforms as long as the nature of the transform remains constant. When any aspect of the transform changes, a new plan needs to be created.

The functions that execute transforms start with `cufftExec`. The main difference between them is the type of the input and output data. Table 8.2 lists the signatures of these functions and provides a description of each.

Table 8.2

Transform Execution

Function	Description
`cufftExecR2C(cufftHandle handle, cufftReal *in, cufftComplex *out)`	Performs the FFT on single-precision real-valued data and returns complex-valued data
`cufftExecC2R(cufftHandle handle, cufftComplex *in, cufftReal *out)`	Performs the FFT on single-precision complex-valued data and returns real-valued data
`cufftExecC2C(cufftHandle handle, cufftComplex *in, cufftComplex *out, int direction)`	Performs the FFT or IFFT on single-precision complex-valued data and returns complex-valued data
`cufftExecD2Z(cufftHandle handle, cufftDoubleReal *in, cufftDoubleComplex *out)`	Performs the FFT on double-precision real-valued data and returns complex-valued data
`cufftExecZ2D(cufftHandle handle, cufftDoubleComplex *in, cufftDoubleReal *out)`	Performs the FFT on double-precision complex-valued data and returns real-valued data
`cufftExecZ2Z(cufftHandle handle, cufftDoubleComplex *in, cufftDoubleComplex *out, int direction)`	Performs the FFT or IFFT on double-precision complex-valued data and returns complex-valued data

In each function, the first argument is the `cufftHandle` configured by the plan function. The next two arguments are pointers to the blocks of device memory that store the input data (`in`) and output data (`out`) for the transform.

The data types of the input/output elements are defined in cufft.h. They resemble the data types introduced in Chapter 6, and the easiest way to understand them is to look at their `typedef` statements:

```
typedef float cufftReal;
typedef double cufftDoubleReal;
typedef cuComplex cufftComplex;
typedef cuDoubleComplex cufftDoubleComplex;
```

As explained in the discussion of cuBLAS, a `cufftComplex` is a `float2` that contains the real and imaginary parts of a complex number. A `cuDoubleComplex` is a `double2` that contains real and imaginary parts. These types are defined in cuComplex.h and they can be accessed using the vector operators discussed in Chapter 3.

If the transform's input and output data are both complex-valued, as in `cufftExecC2C` and `cufftExecZ2Z`, the last argument of the function identifies the desired direction of the transform. This can be set to one of two values:

- `CUFFT_FORWARD` — performs a forward FFT
- `CUFFT_INVERSE` — performs an inverse FFT

The next section presents an example application that transforms data using the forward FFT, and then performs the inverse FFT. Once both steps are complete, it checks the output to make sure it equals the original input.

8.2.3 Example Application

The ch08_fft_ifft project demonstrates how an application can use cuFFT to perform the forward and inverse FFTs on a one-dimensional array containing 128 values. The code is contained in ch08_fft_ifft/main.cu, which performs seven main steps:

1. Calls `cudaMalloc` to allocate blocks of memory on the device and `cudaMemcpy` to initialize the input data with 128 random values between –100 and 100.
2. Calls `cufftPlan1d` to create the plan for the 128-point forward FFT.
3. Calls `cufftExecR2C` to execute the forward FFT from real to complex values.
4. Calls `cufftPlan1d` to create the plan for the 128-point inverse FFT.
5. Calls `cufftExecC2R` to execute the inverse FFT from complex to real values.
6. Calls `cudaMemcpy` to transfer the processed data to the host.
7. Compares the processed data to the original input data and checks if the FFT and IFFT were performed correctly.

Listing 8.1 presents the code in the ch08_fft_ifft/main.cu source file.

Listing 8.1 ch08_fft_ifft/main.cu

```
int main() {
  const int N = 128;

  // Allocate device memory
  int complexSize = N / 2 + 1;
  float *inputMem, *ifftOutputMem;
  cufftComplex* fftOutputMem;
  checkCudaErrors(cudaMalloc(&inputMem, N * sizeof(float)));
  checkCudaErrors(cudaMalloc(&fftOutputMem, complexSize *
    sizeof(cufftComplex)));
  checkCudaErrors(cudaMalloc(&ifftOutputMem, sizeof(float) * N));

  // Fill input with random data
  std::random_device rd;
  std::mt19937 gen(rd());
  std::uniform_real_distribution<float> dist(-100.0f, 100.0f);
  std::vector<float> inputData(N);
  std::generate(inputData.begin(), inputData.end(),
    [&]() { return dist(gen); });
  checkCudaErrors(cudaMemcpy(inputMem, inputData.data(),
    N * sizeof(float), cudaMemcpyHostToDevice));

  // Create plan for forward FFT
  cufftHandle fwdHandle;
  checkcuFftError(cufftPlan1d(&fwdHandle, N, CUFFT_R2C, 1));

  // Execute forward FFT
  checkcuFftError(cufftExecR2C(fwdHandle, inputMem,
    fftOutputMem));

  // Create plan for inverse IFFT
  cufftHandle invHandle;
  checkcuFftError(cufftPlan1d(&invHandle, N, CUFFT_C2R, 1));

  // Execute inverse FFT
  checkcuFftError(cufftExecC2R(invHandle, fftOutputMem,
    ifftOutputMem));

  // Copy result to host
  std::vector<float> outputData(N);
  checkCudaErrors(cudaMemcpy(outputData.data(),
    ifftOutputMem, sizeof(float) * N, cudaMemcpyDeviceToHost));
```

Listing 8.1 ch08_fft_ifft/main.cu (Continued)

```
  // Divide result by N
  for (int i = 0; i < N; i++) {
    outputData[i] /= N;
  }

  // Check error
  float diff, error = 0.0;
  for (size_t i = 0; i<N; i++) {
    diff = inputData[i] - outputData[i];
    error += diff * diff;
  }
  std::cout << "Error: " << error << std::endl;

  // Free resources
  cufftDestroy(fwdHandle);
  cufftDestroy(invHandle);
  cudaFree(inputMem);
  cudaFree(fftOutputMem);
  cudaFree(ifftOutputMem);
  return 0;
}
```

In this code, the FFT converts a real-valued sequence to a complex-valued sequence (R2C). Because the input has no imaginary values, the FFT results can be stored in N/2 + 1 complex values. This explains why the application only allocates 129 `cufftComplex` values for the FFT output (`fftOutputMem`).

After the FFT is performed, the application creates a second plan and calls `cufftExecC2R`. The application doesn't need to specify that the second transformation is an inverse FFT because a forward FFT can't transform complex values into real values. The plan is configured with `CUFFT_C2R`, so the inverse FFT will be performed.

The IFFT doesn't scale the output in the manner described in the first section. For this reason, the application iterates through the IFFT output values and divides each by N. Once this is done, the application computes the error and deallocates resources.

Each cuFFT function called in this code returns a `cufftResult_t` that identifies the status of the operation. Each function call is surrounded by the `checkcuFftError` macro, which accepts the `cufftResult_t` and calls the inline function `checkcuFftErr` to print the file name, line number, and a description of the cuFFT error.

8.3 Summary

Of the mathematical operations discussed in this book, the FFT is my favorite. No other algorithm combines mathematical beauty with practical utility. It's the central algorithm in the world of signal processing, and I've met engineers who have spent their lives struggling to accelerate FFT execution.

The first part of this chapter introduced the Discrete Fourier Transform (DFT), which transforms time-domain samples x(n) into an array of frequency components, X(k). The inverse DFT transforms the frequency components back into a time-domain signal.

The next discussion explains how applications can use the cuFFT library to perform frequency analysis. One unique aspect of cuFFT is the need to create a plan that identifies aspects of the transform including its dimensions, data types, and memory layout. Once an application has created a plan, it can run multiple transforms with the same properties.

The example project for this chapter demonstrates how cuFFT can be used to perform the forward and inverse FFTs. It generates an array of real values and then calls `cufftExecR2C` to compute the forward FFT, which produces an array of complex values. Once this is finished, the application calls `cufftExecC2R` to compute the inverse FFT, which produces an array of real values. When this is done, the application checks the original array against the processed array to verify that the two FFTs produced correct results.

Chapter 9

Processing Images with NPPI

Up to this point, all of our processing has involved general data, usually containing a single dimension. But this chapter explains how to process images. To be specific, the goal is to explain how to process images using the Nvidia Performance Primitives (NPP) package, which consists of three main parts:

- **NPP Core (NPPC)** — provides basic functionality
- **NPP Image Processing (NPPI)** — functions and structures for processing images
- **NPP Signal Processing (NPPS)** — functions and structures for processing signals

This chapter focuses on the data structures and functions provided by NPPI. The first part introduces new data types and functions, and the second part presents functions that perform simple arithmetic operations. This discussion ends by presenting example code that uses these functions to update an image's contrast.

The next part of the chapter explores NPPI's capabilities for image transformation, which include mirroring, rotation, and resizing. This discussion presents example code that uses NPPI's functions to resize and rotate an image.

The final section of the chapter explains how NPPI can be used to perform image filtering. The discussion begins with an explanation of *convolution*, the mathematical operation that makes image filtering possible. Convolution computes a series of 2-D dot products that multiply the pixels of an image by values in a kernel. The nature of the kernel determines the type of filtering to be performed, and the last example project discussed in this chapter demonstrates how NPPI can be used to sharpen an image.

9.1 Overview

The first time I looked at NPPI code, I found it impossible to read. The data types are unlike anything else in CUDA, the function names are bizarre, and there are many sublibraries to keep track of. The goal of this section is to introduce the new data types, the function naming convention, and the sublibraries that make up NPPI.

9.1.1 Data Types

Before you can call NPP or NPPI functions, you need to be familiar with the new data types. This discussion explores the primitive types provided by NPP, the `NppStreamContext` type, and the `NppiSize` and `NppiPoint` types.

Primitive Types

NPP has its own set of numeric data types, and each name starts with `Npp` followed by the number of bits in each instance. This is made clear in Table 9.1, which lists the names of NPP's data types and provides a description of each.

Table 9.1

NPP Primitive Types

NPP Type	Regular Data Type	Description
`Npp8u`	`unsigned char`	8-bit unsigned char
`Npp8s`	`signed char`	8-bit signed char
`Npp16u`	`unsigned short`	16-bit unsigned integer
`Npp16s`	`signed short`	16-bit signed integer
`Npp32u`	`unsigned long`	32-bit unsigned integer
`Npp32s`	`signed long`	32-bit signed integer
`Npp64u`	`unsigned long long`	64-bit unsigned integer
`Npp64s`	`signed long long`	64-bit signed integer
`Npp32f`	`float`	32-bit floating-point value
`Npp64f`	`double`	64-bit floating-point value
`Npp32fc`	`cuComplex`	32-bit floating-point complex value
`Npp64fc`	`cuDoubleComplex`	64-bit floating-point complex value

These types are straightforward to use and understand. For example, the applications in this chapter process images that contain 8-bit unsigned values. Therefore, image data is accessed using `Npp8u` pointers.

NppStreamContext

The first part of Chapter 5 explained that a stream is a queue of commands to be executed by the device. Normally, all commands belong to the default stream, which means they're executed in the order in which they enter the stream. But if an application creates multiple streams, the execution of commands in one stream may overlap with the execution of commands in another stream.

NPPI functions end with `_Ctx`, which means their last argument is a `NppStreamContext`. This has several fields, including the following five:

- `hStream` — the `cudaStream_t` that represents the function's stream
- `nCudaDeviceId` — integer that identifies the target device
- `nMultiProcessorCount` — integer that identifies the number of processors
- `nMaxThreadsPerMultiProcessor` — integer that identifies the maximum threads per multiprocessor
- `nMaxThreadsPerBlock` — identifies the maximum threads per thread block

By accessing these fields, developers can specify which stream, device, and resource state to use when running an image-processing operation on the device. This makes it possible to take advantage of asynchronous execution instead of executing every operation in the default stream.

NppiSize and NppiPoint

Several functions make it possible to specify a region of interest (ROI) inside the image that should be affected by the operation. These ROIs are provided using `NppiSize` instances, and each has a `width` and `height`.

While the `NppiSize` sets the size of the ROI, the location is usually set in one of two ways. If the function accepts an `NppiPoint` argument, the ROI's upper-left corner will be determined by the `NppiPoint`, whose `x` and `y` fields identify pixel coordinates.

If the function doesn't accept an `NppiPoint`, then the upper-left corner of the ROI is set to the upper-left corner of the input image. This is determined by the pointer to the image data.

9.1.2 Function Names

In addition to custom data types, NPP uses a unique naming convention for its functions. For basic NPP functions, each name starts with `npp` followed by the function's operation, as in `nppGetLibVersion`.

For image processing, function names become significantly more complex. The general structure of an NPPI function name is given as follows:

```
nppi<OpName>_<Type>_C<NumChannels><I|R>Suffix
```

The parts of this name are given as follows:

- `OpName` — The operation to be performed (`Add`, `Filter`, and so on)
- `Type` — The data type of each channel in a pixel (`8u`, `16u`, `16s`, `16f`, `32f`)
- `NumChannels` — The number of channels (components) of each pixel
- `I}R` — Identifies if the operation is in-place (`I`) or in a destination region (`R`)
- `Suffix` — Additional information about the function

To understand this, it's important to understand the difference between *pixels* and *channels*. A pixel (picture element) is a distinct point in an image and each pixel can be thought of as a data structure. The fields that make up the structure of each pixel are called the pixel's channels.

Channels commonly contain color information (intensity of red, green, and blue) as well as the degree of transparency (alpha). In a grayscale image, each pixel contains a single value that identifies its luminance. For this reason, we say that grayscale images have a single channel.

To demonstrate how this naming convention works, consider the following function:

```
nppiAdd_8u_C1RSfs_Ctx
```

As given by the `Add` operation name, this adds pixels of two images and produces a third. The `C1` indicates that the input images contain a single channel (grayscale). The `8u` indicates that the data in this channel must be provided as unsigned 8-bit integers. The `R` indicates that the output will be stored in a destination image.

If a function processes integer data, the suffix will contain `Sfs`, which stands for *scale factor shift*. If this is present, the function accepts an integer scale factor that will be used to prevent result values from exceeding the maximum size of the pixel value. If the scale factor is given by s, then each pixel is divided by 2^s after the operation is performed.

For example, suppose an application adds two images whose pixels contain 8-bit unsigned integers (`Npp8u`). The maximum value is 255, and when the pixels are added, the maximum result is 510, which can't be stored in the resulting image. If the scale factor is set to 1, each result will be divided by $2^1 = 2$, which falls into the accepted range.

The function name ends with `_Ctx`, which means the last argument must be set to an `NppStreamContext`. As a result, the function's operation will be inserted into the given stream and executed in sequence with other operations in the stream.

9.1.3 NPPI Libraries

When you build an application that uses NPPI, you can link the entire toolset with a flag like `-lnppi`. This links every NPPI library into the application, which produces a very large result.

NPPI consists of ten libraries and Table 9.2 lists each of them. The second column lists the name of the library file that can be linked into the application.

Table 9.2

NPPI Libraries

Library Name	Library File Name
Arithmetic and Logical Operations	nppial
Color and Sampling Conversion	nppicc
Compression	nppicom
Data Exchange and Initialization	nppidei
Filtering Functions	nppif
Geometry Transforms	nppig
Morphological Transforms	nppim
Linear Transforms/Statistical Operations	nppist
Memory Management	nppisu
Threshold and Compare Operations	nppitc

For example, suppose your application needs NPPI for arithmetic operations and linear transforms. If you're using a GNU linker, you can link the two libraries using the `-lnppial` and `-lnppist` flags. If you use `-lnppi`, the linker will integrate every library in the table.

9.2 Arithmetic Operations

Many common image processing operations can be accomplished using pixel arithmetic. The nppial library provides several functions that perform arithmetic operations, and Table 9.3 lists 21 of them.

Table 9.3

Arithmetic Functions

Function	Description
`Abs(*src, srcStep, *dst, dstStep, roi)`	Compute the absolute value of each pixel in the image
`AbsDiff(*src1, src1Step, *src2, src2Step, *dst, dstStep, roi)`	Compute the absolute difference of the pixels in two images
`AbsDiffC(*src1, src1Step, *dst, dstStep, roi, constant)`	Compute the absolute difference between each pixel in an image and a constant value
`AddC(*src1, src1Step, constant, *dst, dstStep, roi, scale)`	Add a constant value to each pixel in an image
`Add(*src1, src1Step, *src2, src2Step, *dst, dstStep, roi, scale)`	Add the pixels of two images together
`AddSquare(*src, srcStep, *mask, maskStep, *srcDst, srcDstStep, roi)`	Add the square of the pixels of one image to another
`AddWeighted(*src, srcStep, *mask, maskStep, *srcDst, srcDstStep, roi, alpha)`	Add the weighted pixels of one image to another
`AddProduct(*src1, src1Step, *src2, src2Step, *mask, maskStep, *srcDst, srcDstStep, roi)`	Multiplies the pixels of two images together and adds the results to the pixels of a third image
`SubC(*src1, src1Step, constant, *dst, dstStep, roi, scale)`	Subtracts a constant value from each pixel in an image
`Sub(*src1, src1Step, *src2, src2Step, *dst, dstStep, roi, scale)`	Subtracts the pixels of one image from the pixels of another image
`MulC(*src1, src1Step, constant, *dst, dstStep, roi, scale)`	Multiplies each pixel of an image by a constant value
`MulCScale(*src1, src1Step, constant, *dst, dstStep, roi)`	Multiplies each pixel by a constant value and scales the result

`Mul(*src1, src1Step, *src2, src2Step, *dst, dstStep, roi, scale)`	Multiplies the pixels of two images together
`MulScale(*src1, src1Step, *src2, src2Step, *dst, dstStep, roi)`	Multiplies the pixels of two images together and scales the result
`DivC(*src1, src1Step, constant, *dst, dstStep, roi, scale)`	Divide each pixel of an image by a constant value
`Div(*src1, src1Step, *src2, src2Step, *dst, dstStep, roi, scale)`	Divide the pixels of an image by the pixels of another image
`Div_Round(*src1, src1Step, *src2, src2Step, *dst, dstStep, roi, rndMode, scale)`	Divide the pixels of an image by the pixels of another image and rounds the resulting values
`Sqr(*src, srcStep, *dst, dstStep, roi, scale)`	Compute the square of each pixel in an image
`Sqrt(*src, srcStep, *dst, dstStep, roi, scale)`	Compute the square root of each pixel in an image
`Ln(*src, srcStep, *dst, dstStep, roi, scale)`	Compute the natural logarithm of each pixel in an image
`Exp(*src, srcStep, *dst, dstStep, roi, scale)`	Compute the exponential function of each pixel in an image

The first column of this table lists the operation names of the corresponding functions, without the initial `nppi` or the following text. For example, the fourth entry of the table is `AddC`, but NPPI doesn't provide a function by that name. But it provides several functions that perform the `AddC` operation, such as `nppiAddC_8u_C1RSfs_Ctx` and `nppiAddC_16f_C1R_Ctx`.

In the parameter lists, most of the pointers are references to images (`src`, `src1`, `src2`, `dst`). The nature of the pointer depends on the data type of each input channel. For example, the function `nppiAddC_8u_C1RSfs_Ctx` expects pointers that reference memory containing 8-bit unsigned integers. The `nppiAddC_16f_C1R_Ctx` function expects pointers that reference memory containing 16-bit floating-point values.

In addition to the pointer, each image has an integer value that identifies its *step* (`srcStep`, `src1Step`, `src2Step`, `dstStep`). An image's step is the number of bytes between the start of one row of the image and the start of the next row. If there's no padding, the step should be set to the number of pixels per row multiplied by the number of channels per pixel multiplied by the number of bytes per channel.

Many operations don't affect the entire source image or source images, but only a specific region of interest (ROI). This is identified by the `roi` parameter, whose type is `NppiSize`. As discussed earlier, this type has two fields: `width` and `height`.

Many of the functions in the table end with `C`, such as `AddC`, `MulC`, and `DivC`. These functions transform a source image using the value of a constant. For example, `AddC` produces an image whose pixels are obtained by adding a constant to the pixels of the original image. `DivC` produces an image whose pixels are obtained by dividing each pixel of the original image by a constant.

Many functions have a `scale` parameter, which is used to scale the results to prevent overflow. It's important to note that this is only available for functions that accept integer values for pixel channels, such as `nppiMulC_8u_C1IRSfs_Ctx`. The `scale` argument isn't available for functions that accept floating-point values for pixel channels, such as `nppiAddC_16f_C1R_Ctx`.

9.3 Changing Image Contrast

For a grayscale image, contrast is the difference in brightness between light and dark regions. Increasing contrast makes light areas lighter and dark areas darker, leaving gray regions essentially unchanged. Reducing contrast performs the reverse operation.

It helps to view the operation mathematically. If `val` is a pixel in a source image whose values range from 0 to 255, the following equation shows how the updated value, `newval`, is computed when contrast is increased by 50%.

$$newval = 1.5 \cdot (val - 128) + 128$$

The code in the ch09_image_contrast project demonstrates how NPPI can be used to increase the contrast of a grayscale image whose pixel values are 8-bit unsigned integers (`Npp8u`). But before I discuss the code, I need to explain the toolset that the project uses to access image data.

9.3.1 Reading and Writing Images

NPPI provides a wealth of functions for image processing, but it doesn't provide any functions that load image data from files or write image data to files. There are many free libraries that make this possible, including OpenCV and ImageMagick, and though these libraries are powerful, they require installation and linking.

To keep the build process as simple as possible, the code in this book relies on the header-based tools provided by Sean T. Barrett (STB). These tools are provided as regular C/C++ headers (*.h), and you can view the full list of them at https://github.com/nothings/stb.

The example code in this book relies on two headers to access image files:

- **stb_image.h** — provides functions that load pixel data from an image file
- **stb_image_write.h** — provides functions that write pixel data to an image file

To load pixel data from a file, the example code calls `stbi_load`, defined in the stb_image.h header. The function's signature is given as:

```
unsigned char* stbi_load(char const *filename, int *width,
  int *height, int *numChannels, int requiredChannels)
```

For example, the following code reads pixel data from smiley.png into an array of unsigned 8-bit integers:

```
unsigned char* imgData =
  stbi_load("smiley.png", &width, &height, &channels, 1)
```

To write pixel data to a file, the example code in this book relies on the `stbi_write_png` function, which is defined in the stb_image_write.h header:

```
int stbi_write_png(char const *filename, int width, int height,
  int numChannels, const void *data, int stride)
```

The fifth argument provides pixel data in an array of 8-bit unsigned integer values. If the image contains P pixels and each pixel has C channels, the array must contain PC elements. The last argument of `stbi_write_png` accepts the number of bytes between the start of one row of pixels and the start of the next row, so the terms *stride* and *step* are synonyms.

Suppose you want to write data from an array named `pixel_array` to an image file named new_smiley.png. If the image's width is `width`, the height is `height`, and the number of channels is `num_channels`, the following code shows how `stbi_write_png` can be used:

```
stbi_write_png("new_smiley.png", width, height, num_channels,
  pixel_array, step)
```

The following discussion presents example code that demonstrates how the STB functions can be used in practice. The application reads from a PNG file with `stbi_load`, processes the image data, and calls `stbi_write_png` to write the processed data to a new PNG file.

9.3.2 Example Application

Now that you understand NPPI's arithmetic functions and the STB functions that access image files, you're ready to explore the ch09_image_contrast project. The goal is to increase the contrast of smiley.png by 50%. As presented earlier, this is accomplished with the following equation:

$$newval = 1.5 \cdot (val - 128) + 128$$

We can reduce the number of operations with the following simplification:

$$\begin{aligned} newval &= 1.5 \cdot val - 1.5 \cdot 128 + 128 \\ &= 1.5 \cdot val - 1.5 \cdot 128 + 128 \\ &= 1.5 \cdot val - 0.5 \cdot 128 = 1.5 \cdot val - 64 \end{aligned}$$

But there's a problem. NPPI doesn't support multiplying integers by floating-point values like 1.5. Instead, it supports multiplication by an integer and then scaling by a power of 2. We can reach essentially the same result by multiplying the value by 6 and then dividing by 4 (2^2). This leads to the final equation:

$$newval = \frac{6 \cdot val}{2^2} - 64$$

To increase contrast in code, the project's main.cu file performs nine steps:

1. It creates an `NppStreamContext` to serve as the stream containing all of the image processing commands.
2. It calls `stbi_load` to read pixel data from smiley.png into an array of 8-bit integers.
3. It allocates a block of memory on the device with the same size as the pixel data.
4. It calls `cudaMemcpyAsync` to copy the pixel data to the allocated device memory.
5. It calls `nppiMulC_8u_C1IRSfs_Ctx` to multiply each pixel by 6 and then scale the results by 2^2.
6. It calls `nppiSubC_8u_C1IRSfs_Ctx` to subtract 64 from each pixel.
7. It calls `cudaMemcpyAsync` to copy processed pixel data to the host.
8. It calls `cudaStreamSynchronize` to wait for the stream's commands to finish.
9. It calls `stbi_write_png` to write the pixel data to a file named new_smiley.png.

Listing 9.1 presents the code in ch09_image_contrast/main.cu. As shown, all of the NPPI operations are performed inside a custom `NppStreamContext`.

Listing 9.1 ch09_image_contrast/main.cu

```
int main() {

  // Get device identifier
  int device;
  checkCudaErrors(cudaGetDevice(&device));

  // Get device properties
  cudaDeviceProp deviceProp;
  checkCudaErrors(cudaGetDeviceProperties(&deviceProp, device));

  // Create stream
  cudaStream_t stream;
  cudaStreamCreate(&stream);

  // Define NPPI stream
  NppStreamContext ctx;
  ctx.hStream = stream;
  ctx.nCudaDeviceId = device;
  ctx.nMultiProcessorCount = deviceProp.multiProcessorCount;
  ctx.nMaxThreadsPerMultiProcessor =
    deviceProp.maxThreadsPerMultiProcessor;
  ctx.nMaxThreadsPerBlock = deviceProp.maxThreadsPerBlock;
  ctx.nSharedMemPerBlock = deviceProp.sharedMemPerBlock;
  ctx.nCudaDevAttrComputeCapabilityMajor = deviceProp.major;
  ctx.nCudaDevAttrComputeCapabilityMinor = deviceProp.minor;

  // Read data from image file
  int width, height, channels;
  unsigned char* imgData = stbi_load("smiley.png",
    &width, &height, &channels, 1);
  if (!imgData) {
    std::cerr << "Error: failed to load image" << std::endl;
    return -1;
  }

  // Allocate memory on the device
  unsigned char* imgMem;
  checkCudaErrors(cudaMallocAsync(
    reinterpret_cast<void**>(&imgMem), width * height,
    ctx.hStream));
```

Listing 9.1 ch09_image_contrast/main.cu (Continued)

```
    // Copy image data to the device
    checkCudaErrors(cudaMemcpyAsync(imgMem, imgData,
      width * height, cudaMemcpyHostToDevice, ctx.hStream));

    // Multiply by scaled contrast
    int step = width;
    NppiSize roi = { width, height };
    checkNppError(nppiMulC_8u_C1IRSfs_Ctx(6, imgMem,
      step, roi, 2, ctx));

    // Subtract 64 from each pixel
    checkNppError(nppiSubC_8u_C1IRSfs_Ctx(64, imgMem,
      step, roi, 0, ctx));

    // Copy image data to the host
    checkCudaErrors(cudaMemcpyAsync(imgData, imgMem,
      width * height, cudaMemcpyDeviceToHost, ctx.hStream));

    // Wait for stream to finish
    cudaStreamSynchronize(ctx.hStream);

    // Write data to new image file
    stbi_write_png("new_smiley.png", width, height,
      1, imgData, step);

    // Free resources
    cudaFreeAsync(imgMem, ctx.hStream);
    cudaStreamDestroy(ctx.hStream);
    stbi_image_free(imgData);
    return 0;
}
```

In earlier versions of CUDA, NPPI provided functions that didn't require an `NppStreamContext`. But the functions in the current version of NPPI do require `NppStreamContexts`, and this is why the code starts by declaring a stream context (`ctx`) and setting its fields.

To ensure sequential operation, all the data operations need to be associated with the same stream context. This explains why the application calls `cudaMallocAsync`, `cudaMemcpyAsync`, and `cudaFreeAsync` instead of `cudaMalloc`, `cudaMemcpy`, and `cudaFree`. After the operations have been inserted into the stream, the application calls `cudaStreamSynchronize` to wait for the operations to finish.

It's important to understand the parameters of `nppiMulC_8u_C1IRSfs_Ctx`. The goal is to multiply each pixel by 1.5, so the first parameter is set to 6 and the fifth parameter is set to 2 ($1.5 = 6/2^2$). Because the image is grayscale, each pixel requires one byte, so the number of bytes in each row (the step) equals the image's width. The operation should affect the entire image, so the region of interest is set to an `NppiSize` whose dimensions equal those of the image.

The application calls `cudaStreamSynchronize` to wait until the NPPI operations complete. Then it calls `stbi_write_png` to write the image data to a file named new_smiley.png. Figure 9.1 shows what this looks like.

Figure 9.1 Updating Image Contrast

The ch09_image_contrast project provides a simple demonstration of NPPI, but for real-world applications, it's a good idea to convert the 8-bit pixels to floating-point values. This avoids issues related to overflow and underflow.

9.4 Geometry Transformation

NPPI's nppig sublibrary provides several functions that move pixels of an image without changing their values. These operations are referred to as *geometry transformations*, and the functions are provided in the nppi_geometry_transforms.h header file.

This section presents three of the most important types of transformations:

- **mirroring** — flips an image vertically, horizontally, or both
- **rotation** — rotates an image around an axis by a specified angle
- **resizing** — shrinks or magnifies the image

9.4.1 Mirroring

The simplest of the geometric transformations involves flipping an image. NPPI provides several functions that accomplish this, and all of their names start with nppiMirror. For example, the function that flips a grayscale image containing 8-bit unsigned values is given as follows:

```
nppiMirror_8u_C1R_Ctx(const Npp8u* src, int srcStep,
  Npp8u *dst, int dstStep, NppiSize roi, NppiAxis flip,
  NppStreamContext nppStreamCtx)
```

Most of these arguments should look familiar. The src and dst parameters point to the blocks of device memory containing the input and output image data. The srcStep and dstStep parameters identify the step of the input and output images. The roi argument identifies the region of interest that should be affected by the operation.

The flip argument requires explanation. This argument identifies the axis that should be used for flipping, and it can be set to one of three values:

- NPP_HORIZONTAL_AXIS — Flip the image horizontally
- NPP_VERTICAL_AXIS — Flip the image vertically
- NPP_BOTH_AXIS — Flip the image horizontally and vertically

The last argument of the function identifies the stream context that will contain the operation's command. As discussed earlier, the NppStreamContext structure is essentially a wrapper around the cudaStream_t structure introduced in Chapter 5.

9.4.2 Rotation

Though conceptually simple, rotating an image with NPPI is significantly more complicated than mirroring an image. NPPI's rotation functions start with nppiRotate, and the following function rotates a grayscale image containing 8-bit unsigned values:

```
nppiRotate_8u_C1R_Ctx(const Npp8u *src, NppiSize srcSize,
  int nSrcStep, NppiRect srcROI, Npp8u *dst, int nDstStep,
  NppiRect dstROI, double angle, double shiftX, double shiftY,
  int eInterpolation, NppStreamContext nppStreamCtx)
```

It's important to understand the difference between the srcSize and srcROI arguments. srcSize identifies the size of the input image and srcROI identifies which region of the input image should be rotated.

When I first tried calling `nppiRotate` in an application, I set the size of the output region of interest equal to the size of the input region of interest. But this causes an error. The rotated image usually takes up more space than the original image, so the destination ROI must be larger. The following discussion explains this further.

Rotation Bounds

Rotating an image with `nppiRotate` produces a larger image than the original. To compute the minimum bounds of the output image, NPPI provides a helper function called `nppiGetRotateBound`:

```
nppiGetRotateBound(NppiRect srcROI, double boundingBox[2][2],
  double angle, double shiftX, double shiftY)
```

This accepts the parameters of the rotation and returns a bounding box that identifies the size of the rotation result. The angle of rotation is given by the `angle` parameter, which should be provided in degrees.

The `shiftX` and `shiftY` parameters deserve explanation. Some sources say that these values set the center of the rotation, but this is not the case. The center of rotation is *always* the upper left corner. `shiftX` and `shiftY` identify how many pixels the rotated image should be shifted *after rotation*.

Interpolation

During rotation, the function needs to estimate values of pixels between known pixels. This is called *interpolation*, and the second-to-last argument of `nppiRotate` identifies the interpolation method to be used. This must be set to one of the values listed in Table 9.4.

Table 9.4

Interpolation Methods

Interpolation Method Name	Description
NPPI_INTER_NN	Estimate pixels using nearest neighbors
NPPI_INTER_LINEAR	Estimate pixels using bilinear interpolation
NPPI_INTER_CUBIC	Estimate pixels using bicubic interpolation
NPPI_INTER_CUBIC2P_BSPLINE	Bicubic interpolation using a B-spline
NPPI_INTER_CUBIC2P_CATMULLROM	Bicubic interpolation using a Catmull-Rom spline
NPPI_INTER_CUBIC2P_B05C03	Interpolation using the Mitchell–Netravali filter

When it comes to image processing, it's important to choose the right interpolation method. The simplest and fastest method is the first, NPPI_INTER_NN, which sets the value of an unknown pixel to the value of the pixels nearest to it. In other words, if Pixel P is closer to Pixel A than any other pixel, interpolation will set the color of Pixel P to the color of Pixel A.

If this method is used to enlarge an image, the resulting image will contain only the colors in the original. For example, if an image is enlarged threefold, each color will be repeated three times. This is shown in Figure 9.2, which depicts four adjacent pixels repeated three times.

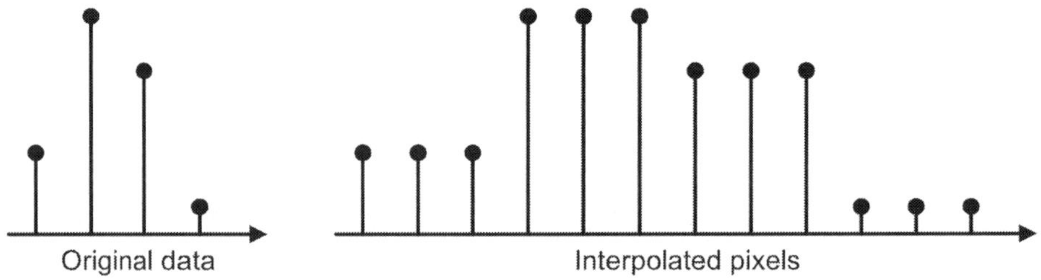

Figure 9.2 Nearest-Neighbor Interpolation

Nearest-neighbor interpolation executes quickly because it doesn't require any math. But when used for image processing, results tend to look grainy and pixellated.

We can improve on this by using bilinear interpolation, which can be configured in CUDA by setting the eInterpolation parameter to setting NPPI_INTER_LINEAR. If this is selected, if a point lies between pixels, the interpolated color will be set equal to a linear combination of the pixels' colors.

To understand this, it helps to grasp the one-dimensional case. Suppose Point P lies on a line between adjacent pixels A and B. To identify the location of P relative to A and B, we use a parameter called t. The distance from P to the center of A is given by t and the distance from P to the center of B is given by 1 – t. This is called linear interpolation, and it sets the color at P with the following equation:

$$P_{color} = t \cdot A_{color} + (1 - t) \cdot B_{color}$$

From this equation, it's clear that if t = 1, the color at P equals the color at A. If t = 0, the color at P equals the color at B. If t = 0.5, the resulting color will be an average of the color of A and the color of B. The resulting color value from A to B follows a straight line, and Figure 9.3 depicts the lines used to interpolate values between four pixels. Note the difference between this data expansion and that shown in Figure 9.2.

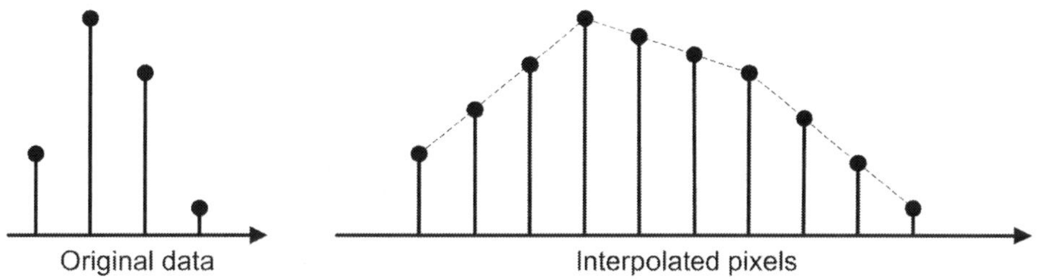

Figure 9.3 Linear Interpolation

In two dimensions, this method is called *bilinear interpolation*. Now we assume that Point P lies between four adjacent pixels: A, B, C, and D. With this arrangement, each point needs two interpolation parameters, t_1 and t_2. Figure 9.4 illustrates how the value of Point P can be obtained using bilinear interpolation.

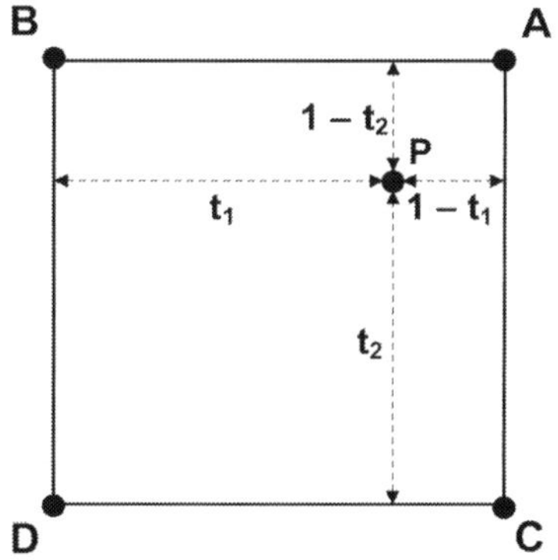

Figure 9.4 Pixels and Points in Bilinear Interpolation

To compute the color at P, bilinear interpolation extends the preceding equation into two dimensions. The new equation is given as follows:

$$P_{color} = t_1 \cdot t_2 \cdot A_{color} + (1 - t_1) \cdot t_2 \cdot B_{color}$$
$$+ t_1 \cdot (1 - t_2) \cdot C_{color} + (1 - t_1) \cdot (1 - t_2) \cdot D_{color}$$

Bicubic interpolation arrives at a similar equation, but connects points using a cubic polynomial instead of a straight line. Cubic interpolation can be configured in NPPI by setting the `eInterpolation` parameter to setting `NPPI_INTER_CUBIC`.

9.4.3 Resizing

Shrinking and enlarging images are common operations in image processing, and NPPI provides several functions that make this possible. The function names start with `nppiResize`, and the function that resizes grayscale images containing 8-bit unsigned values is `nppiResizeSqrPixel_8u_C1R_Ctx`:

```
nppiResizeSqrPixel_8u_C1R_Ctx(const Npp8u *src,
  NppiSize srcSize, int srcStep, NppiRect srcROI,
  Npp8u *dst, int dstStep, NppiRect dstROI, double xFactor,
  double yFactor, double xShift, double yShift,
  int eInterpolation, NppStreamContext nppStreamCtx)
```

Most of these parameters are identical to those in the `nppiRotate_8u_C1R_Ctx` function presented earlier. The only new arguments are `xFactor` and `yFactor`, which identify how much the image should be resized in the x-direction and y-direction. If one of these is less than 1, the image will be reduced in the corresponding direction. If a value is greater than 1, the image will be enlarged in that direction.

9.4.4 Example Application

The ch09_rotate_resize project demonstrates how NPPI can be used to rotate and resize a grayscale image. The code in the ch09_rotate_resize performs twelve operations:

1. It creates an `NppStreamContext` to serve as the application's stream context.
2. It calls `stbi_load` to read pixel data from smiley.png into an array of 8-bit integers.
3. It allocates a block of memory on the device to hold the image data.
4. It calls `cudaMemcpyAsync` to copy the image data to the allocated device memory.
5. It calls `nppiGetRotateBound` to obtain the bounds of the rotation result.
6. It allocates a block of memory on the device to store the rotation result.
7. It calls `nppiRotate_8u_C1R_Ctx` to rotate the input image.
8. It allocates a block of memory on the device to store the resize result.
9. It calls `nppiResizeSqrPixel_8u_C1R_Ctx` to resize the rotated image.
10. It calls `cudaMemcpyAsync` to copy the rotated/resized image data to the host.
11. It calls `cudaStreamSynchronize` to wait for the stream's commands to finish.
12. It calls `stbi_write_png` to write the pixel data to a new image file named new_smiley.png.

Listing 9.2 presents the code in ch09_rotate_resize/main.cu. Rotation and resizing can't be performed in-place, so the application allocates three blocks of device memory to hold pixel data: `imgMem1`, `imgMem2`, and `imgMem3`.

Listing 9.2 ch09_rotate_resize/main.cu

```
int main() {

  // Get device identifier
  int device;
  checkCudaErrors(cudaGetDevice(&device));
  cudaDeviceProp deviceProp;
  checkCudaErrors(cudaGetDeviceProperties(&deviceProp, device));

  // Create stream
  cudaStream_t stream;
  cudaStreamCreate(&stream);

  // Define NPPI stream
  NppStreamContext ctx;
  ctx.hStream = stream;
  ctx.nCudaDeviceId = device;
  ctx.nMultiProcessorCount = deviceProp.multiProcessorCount;
  ctx.nMaxThreadsPerMultiProcessor =
    deviceProp.maxThreadsPerMultiProcessor;
  ctx.nMaxThreadsPerBlock = deviceProp.maxThreadsPerBlock;
  ctx.nSharedMemPerBlock = deviceProp.sharedMemPerBlock;
  ctx.nCudaDevAttrComputeCapabilityMajor = deviceProp.major;
  ctx.nCudaDevAttrComputeCapabilityMinor = deviceProp.minor;

  // Read data from image file
  int width, height, channels;
  unsigned char* inData =
    stbi_load("smiley.png", &width, &height, &channels, 1);
  if (!inData) {
    std::cerr << "Error: failed to load image" << std::endl;
    return -1;
  }

  // Allocate memory on the device
  unsigned char *imgMem1;
  checkCudaErrors(cudaMallocAsync(
    reinterpret_cast<void**>(&inData), width * height,
    ctx.hStream));
```

Listing 9.2 ch09_rotate_resize/main.cu (Continued)

```
// Copy image data to the device
checkCudaErrors(cudaMemcpyAsync(imgMem1, imgData1,
  width * height, cudaMemcpyHostToDevice, ctx.hStream));

// Get the bounds of rotation
double boundingBox[2][2];
NppiRect srcRoi = { 0, 0, width, height };
float shiftX = width / 2.0f, shiftY = height / 2.0f;
checkNppError(nppiGetRotateBound(srcRoi, boundingBox,
  45.0, shiftX, shiftY));

// Allocate memory for rotated image
unsigned char* imgMem2;
int dstWidth = static_cast<int>(ceil(boundingBox[1][0] -
  boundingBox[0][0]));
int dstHeight = static_cast<int>(ceil(boundingBox[1][1] -
  boundingBox[0][1]));
checkCudaErrors(cudaMallocAsync(
  reinterpret_cast<void**>(&imgMem2), dstWidth * dstHeight,
  ctx.hStream));

// Rotate image
int srcStep = width, dstStep = dstWidth;
double sy = ceil(dstHeight / 2.0);
NppiSize size = { width, height };
NppiRect dstRoi = { 0, 0, dstWidth, dstHeight };
checkNppError(nppiRotate_8u_C1R_Ctx(imgMem1, size,
  srcStep, srcRoi, imgMem2, dstStep, dstRoi, 45.0, 0.0,
  sy, NPPI_INTER_CUBIC, ctx));

// Allocate memory for rotated image
unsigned char* imgMem3;
double xFactor = 2.0, yFactor = 2.0;
int outWidth = dstWidth * static_cast<int>(xFactor);
int outHeight = dstHeight * static_cast<int>(yFactor);
checkCudaErrors(cudaMallocAsync(
  reinterpret_cast<void**>(&imgMem3), outWidth * outHeight,
  ctx.hStream));

// Resize image
size = { dstWidth, dstHeight };
int outStep = outWidth;
NppiRect outRoi = { 0, 0, outWidth, outHeight };
```

Listing 9.2 ch09_rotate_resize/main.cu (Continued)

```
  checkNppError(nppiResizeSqrPixel_8u_C1R_Ctx(imgMem2,
    size, dstStep, dstRoi, imgMem3, outStep, outRoi,
    xFactor, yFactor, 0, 2, NPPI_INTER_CUBIC, ctx));

  // Copy image data to the host
  unsigned char* outData =
    (unsigned char*)malloc(outWidth * outHeight);
  checkCudaErrors(cudaMemcpyAsync(outData, imgMem3,
    outWidth * outHeight, cudaMemcpyDeviceToHost, ctx.hStream));

  // Wait for stream to finish
  cudaStreamSynchronize(ctx.hStream);

  // Write data to new image file
  stbi_write_png("new_smiley.png", outWidth, outHeight, 1,
    outData, outStep);

  // Free resources
  cudaFreeAsync(imgMem1, ctx.hStream);
  cudaFreeAsync(imgMem2, ctx.hStream);
  cudaFreeAsync(imgMem3, ctx.hStream);
  cudaStreamDestroy(ctx.hStream);
  stbi_image_free(inData);
  free(outData);
  return 0;
}
```

The application needs to call `nppiGetRotateBound` to get the dimensions of the rotated image. In this case, the rotated image occupies 347-by-347 pixels. The `xFactor` and `yFactor` parameters of the resize function are both set to 2.0, so the resized image has twice the size of the rotated image: 694-by-694 pixels.

In `nppiRotate_8u_C1R_Ctx`, the `shiftX` parameter is set to 0 and the `shiftY` parameter is set to half of the height. As a result, the rotated image will be shifted downward by half the height. This is necessary to make the rotation appear as though it was centered at the image's center.

In the rotation function, `nppiRotate_8u_C1R_Ctx`, and the resize function, `nppiResizeSqrPixel_8u_C1R_Ctx`, the interpolation method is set to `NPPI_INTER_CUBIC`. This tells NPPI to compute pixel values using bicubic interpolation, which creates a cubic polynomial between known pixels and assigns values to points on the curve. On the following page, Figure 9.5 shows what the resulting images look like.

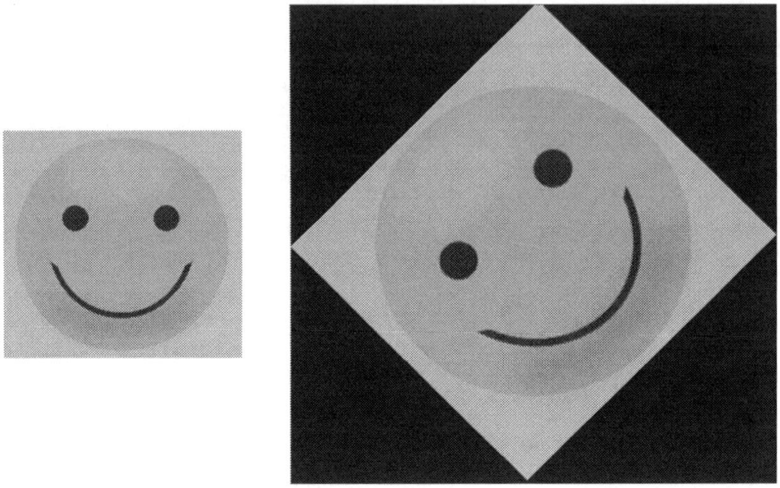

Figure 9.5 Image Rotation and Resizing

As in the preceding example, all of the NPPI operations are performed in the same `NppStreamContext`. This has an `hStream` field that must be set to a `cudaStream_t` structure, which was discussed in Chapter 5. Every data transfer operation is directed to this stream to ensure that they're performed in the order given in the application.

The application calls `cudaStreamSynchronize` with the context's stream to wait for these operations to complete. Then it calls `cudaFreeAsync` to free the blocks of device memory, `cudaStreamDestroy` to deallocate the stream structure, and `stbi_image_free` to free the array of pixels read from the input file.

9.5 Filtering

If you've used image editing applications like Adobe Photoshop, you're probably familiar with image filtering, which adds effects to images such as blurring, sharpening, or embossing. At a low level, these tools perform their operations using a mathematical process called *convolution*.

Convolution plays a critical role in image filtering, and the first part of this section explains how it works. Then we'll explore the NPPI functions that make it possible to perform convolution at high speed on Nvidia GPUs. This section ends with a practical example that demonstrates how NPPI's functions make it possible to filter images.

9.5.1 Convolution

Image convolution replaces each pixel of an image with the result of a two-dimensional dot product. This dot product accepts two matrices and returns the sum of the products of their corresponding elements. For example, suppose A and B are two 3-by-3 matrices whose elements are given as follows:

$$A = \begin{vmatrix} 1 & 2 & 3 \\ 4 & 5 & 6 \\ 7 & 8 & 9 \end{vmatrix} \qquad B = \begin{vmatrix} -10 & -9 & -8 \\ -7 & -6 & -5 \\ -4 & -3 & -2 \end{vmatrix}$$

The two-dimensional dot product of A and B is computed by multiplying corresponding pairs of values and adding the results together:

$$1(-10)+2(-9)+3(-8)+4(-7)+5(-6)+$$
$$6(-5)+7(-4)+8(-3)+9(-2)=-210$$

The first matrix involved in the convolution is the M-by-N rectangle surrounding one of the image's pixels. The second matrix is commonly called a *kernel* or a *filter*. The elements of the filter determine what effect it will have on the image.

For example, if part of an image is denoted as a matrix M, the pixel in the ith row and jth column is $m_{i,j}$. If the filter is denoted as a matrix K, the element in the ith row and jth column is $k_{i,j}$. With this notation, the convolution process obtains the new value of $m_{i,j}$ with the following dot product:

$$m_{i,j}(filtered) = \begin{vmatrix} m_{i-1,j-1} & m_{i-1,j} & m_{i-1,j+1} \\ m_{i,j-1} & m_{i,j} & m_{i,j+1} \\ m_{i+1,j-1} & m_{i+1,j} & m_{i-1,j+1} \end{vmatrix} \cdot \begin{vmatrix} k_{0,0} & k_{0,1} & k_{0,2} \\ k_{1,0} & k_{1,1} & k_{1,2} \\ k_{2,0} & k_{2,1} & k_{2,2} \end{vmatrix}$$
$$= m_{i-1,j-1}k_{0,0} + m_{i-1,j}k_{0,1} + m_{i-1,j+1}k_{0,2} + m_{i,j-1}k_{1,0} + m_{i,j}k_{1,1}$$
$$+ m_{i,j+1}k_{1,2} + m_{i+1,j-1}k_{2,0} + m_{i+1,j}k_{2,1} + m_{i-1,j+1}k_{2,2}$$

When filtering an image, this dot product needs to be computed for every pixel. This can take a significant amount of time, so for large images, some toolsets perform convolution using the Fast Fourier Transform (FFT) discussed in Chapter 8.

A good way to understand image filtering is to look at an example. The grayscale image illustrated in Figure 9.6 has a great deal of noise.

Figure 9.6 Image with Noise

Mathematically, noise can be thought of as unwanted variation between adjacent pixels. This variation can be reduced by replacing each pixel with the average of itself and its surrounding pixels. This averaging operation can be accomplished by convolving the image with a kernel like the following:

$$k = \begin{vmatrix} \frac{1}{9} & \frac{1}{9} & \frac{1}{9} \\ \frac{1}{9} & \frac{1}{9} & \frac{1}{9} \\ \frac{1}{9} & \frac{1}{9} & \frac{1}{9} \end{vmatrix}$$

Denoting the pixel in the ith row and jth column as m_{ij}, the pixels in the filtered image can be computed with the following equation:

$$m_{i,j}(filtered) = \frac{1}{9}m_{i-1,j-1} + \frac{1}{9}m_{i-1,j} + \frac{1}{9}m_{i-1,j+1} + \frac{1}{9}m_{i,j-1}$$
$$+\frac{1}{9}m_{i,j} + \frac{1}{9}m_{i,j+1} + \frac{1}{9}m_{i+1,j-1} + \frac{1}{9}m_{i+1,j} + \frac{1}{9}m_{i+1,j+1}$$

This type of filter is called an *averaging filter* (or a *box filter*) because the pixels of the resulting image are computed by taking the average of the pixels in the original image. After convolution, the filtered pixels form the image illustrated in Figure 9.7.

Figure 9.7 Image after Filtering

In addition to removing noise, the averaging filter removed a lot of detail that isn't noise. To improve on this, engineers frequently use a more effective noise-reduction filter called the *Gaussian filter*. Instead of setting each value in the kernel to a constant, the elements in a Gaussian filter are set equal to values of the Gaussian curve. Chapter 12 will provide a thorough discussion of Gaussian filters.

9.5.2 Filtering Function

NPPI provides several functions for image filtering in the nppif sublibrary, and they're declared in the nppi_filtering_functions.h header file. Some functions perform convolution by column (`nppiFilterColumn`) while others perform convolution by row (`nppiFilterRow`).

For two-dimensional convolution, there are functions that perform box filtering (`nppiFilterBox`), maximum filtering (`nppiFilterMax`), and median filtering (`nppiFilterMedian`). The most general filtering function is `nppiFilter`, and the specific version of the function that accepts 8-bit unsigned grayscale pixels has the following signature:

```
nppiFilter_8u_C1R_Ctx(const Npp8u* src, Npp32s srcStep,
  Npp8u *dst, Npp32s dstStep, NppiSize sizeROI,
  const Npp32s *kernel, NppiSize kernelSize, NppiPoint anchor,
  Npp32s divisor, NppStreamContext nppStreamCtx)
```

Most of these arguments should look familiar. `src` points to the block of device memory containing the input pixels and `dst` points to the device memory that should contain the filtered result. `srcStep` identifies the number of bytes between successive rows in the input and `dstStep` identifies the number of bytes between successive rows in the output.

Four parameters of this function relate to the kernel that should be used for filtering, and they're given as:

1. `kernel` — block of device memory that contains the kernel coefficients
2. `kernelSize` — number of coefficients in the kernel
3. `anchor` — index of the central coefficient in the kernel
4. `divisor` — value by which the filtered results should be divided

For example, the preceding discussion explained that the kernel of a simple averaging filter has nine coefficients, and they all equal 1/9. To use this kernel for filtering, you'd set `kernel` to an array of nine integers all set to 1, `kernelSize` to (3, 3), `anchor` to (1, 1), and `divisor` to 9.

It's important to notice that the `kernel` parameter can only contain signed integers, not floating-point values. Therefore, to define a filter like the averaging filter, you'll need to set `divisor` to a value that each filtered pixel will be divided by. In the example averaging filter, the filtered values need to be divided by 9, so `divisor` should be set to 9.

In the preceding example, the kernel was given as a 3-by-3 matrix. Kernels are usually set to square matrices, and if the size is 5-by-5, then `kernelSize` would be set to (5, 5) and `anchor` would be set to (2, 2). This is because the coefficient in the third row and third column is the center of a 5-by-5 matrix.

When performing convolution, an important question arises: How should the two-dimensional product be computed for pixels on the image's border? That is, what values should be assigned to the surrounding pixels? The `NppiFilter` functions don't address this, so NPPI provides another set of functions that start with `NppiFilterBorder`:

```
nppiFilterBorder_8u_C1R_Ctx(const Npp8u* src, Npp32s srcStep,
  NppiSize srcSize, NppiPoint srcOffset, Npp8u *dst,
  Npp32s dstStep, NppiSize sizeROI, const Npp32s *kernel,
  NppiSize kernelSize, NppiPoint anchor, Npp32s divisor,
  NppiBorderType borderType, NppStreamContext nppStreamCtx)
```

The `borderType` argument tells NPPI how to assign values to pixels outside the image's border. This can be set to one of five values:

- `NPP_BORDER_NONE` — No border handling is performed
- `NPP_BORDER_CONSTANT` — Sets out-of-bounds pixels to a constant value (0)
- `NPP_BORDER_REPLICATE` — Sets out-of-bounds pixels to the nearest valid value
- `NPP_BORDER_WRAP` — The image is treated as though the pixels wrap around
- `NPP_BORDER_MIRROR` — Pixels are reflected at the boundary

9.5.3 Image Sharpening

While a box filter is used to reduce the difference between pixels, a *sharpening filter* is used to accentuate the difference. That is, instead of adding values of neighboring pixels, a sharpening filter subtracts them from the central pixel. An example sharpening kernel is shown here:

$$k = \begin{vmatrix} 0 & -1 & 0 \\ -1 & 5 & -1 \\ 0 & -1 & 0 \end{vmatrix}$$

The central pixel has a much greater value than its neighbors, so the filtered value will be significantly magnified. This is shown in Figure 9.8, which depicts the original image on the left and the sharpened image on the right.

Figure 9.8 Effect of a Sharpening Filter

The ch09_image_filter project demonstrates how NPPI can be used to implement a sharpening filter. The code in main.cu performs ten operations:

1. It creates an `NppStreamContext` to serve as the application's stream context.
2. It calls `stbi_load` to read pixel data from smiley.png into an array of 8-bit integers.
3. It allocates a block of memory on the device to hold the input image and another block to hold the filtered image.
4. It calls `cudaMemcpyAsync` to copy the image data to the allocated block of device memory.

5. It defines a kernel array on the host that contains coefficients of the sharpening filter.
6. It allocates a block of device memory for the kernel and calls `cudaMemcpyAsync` to copy the kernel coefficients into the block.
7. It calls `nppiFilter_8u_C1R_Ctx` to perform filtering with the specified kernel.
8. It calls `cudaMemcpyAsync` to copy the filtered image data to the host.
9. It calls `cudaStreamSynchronize` to wait for the stream's commands to finish.
10. It calls `stbi_write_png` to write the pixel data to a new image file named new_smiley.png.

Listing 9.3 presents the code in ch09_image_filter/main.cu. At the start, it creates a `cudaStream_t` and then an `NppStreamContext`. Then it uses these structures throughout the program to synchronize data operations and the execution of the image filtering.

Listing 9.3 ch09_image_filter/main.cu

```
int main() {

  // Get device identifier
  int device;
  checkCudaErrors(cudaGetDevice(&device));

  // Get device properties
  cudaDeviceProp deviceProp;
  checkCudaErrors(cudaGetDeviceProperties(&deviceProp, device));

  // Create stream
  cudaStream_t stream;
  cudaStreamCreate(&stream);

  // Define NPPI stream
  NppStreamContext ctx;
  ctx.hStream = stream;
  ctx.nCudaDeviceId = device;
  ctx.nMultiProcessorCount = deviceProp.multiProcessorCount;
  ctx.nMaxThreadsPerMultiProcessor =
    deviceProp.maxThreadsPerMultiProcessor;
  ctx.nMaxThreadsPerBlock = deviceProp.maxThreadsPerBlock;
  ctx.nSharedMemPerBlock = deviceProp.sharedMemPerBlock;
  ctx.nCudaDevAttrComputeCapabilityMajor = deviceProp.major;
  ctx.nCudaDevAttrComputeCapabilityMinor = deviceProp.minor;
```

Listing 9.3 ch09_image_filter/main.cu (Continued)

```
// Define kernel of sharpening filter
Npp32s kernel[9] = { 0, -1, 0, -1, 5, -1, 0, -1, 0 };
NppiSize kernelSize = { 3, 3 };
NppiPoint anchor = { 1, 1 };

// Read data from image file
int width, height, channels;
unsigned char* imgData = stbi_load("smiley.png", &width,
  &height, &channels, 1);
if (!imgData) {
  std::cerr << "Error: failed to load image" << std::endl;
  return -1;
}

// Allocate memory on the device
unsigned char *inMem, *outMem;
checkCudaErrors(cudaMallocAsync(
  reinterpret_cast<void**>(&inMem), width * height,
  ctx.hStream));
checkCudaErrors(cudaMallocAsync(
  reinterpret_cast<void**>(&outMem), width * height,
  ctx.hStream));

// Copy image data to the device
checkCudaErrors(cudaMemcpyAsync(inMem, imgData,
  width * height, cudaMemcpyHostToDevice, ctx.hStream));

// Define kernel of sharpening filter
Npp32s kernel[9] = { 0, -1, 0, -1, 5, -1, 0, -1, 0 };
NppiSize kernelSize = { 3, 3 };
NppiPoint anchor = { 1, 1 };

// Allocate device memory for the kernel
Npp32s *kernelMem;
checkCudaErrors(cudaMallocAsync(
  reinterpret_cast<void**>(&kernelMem), 9 * sizeof(Npp32s),
  ctx.hStream));

// Copy the kernel to the device
checkCudaErrors(cudaMemcpyAsync(kernelMem, kernel,
  9 * sizeof(Npp32s), cudaMemcpyHostToDevice, ctx.hStream));
```

Listing 9.3 ch09_image_filter/main.cu (Continued)

```
    // Filter image
    int step = width;
    NppiSize size = { width, height };
    NppiPoint offset = { 0, 0 };
    checkNppError(nppiFilterBorder_8u_C1R_Ctx(inMem, step,
      size, offset, outMem, step, size, kernelMem, kernelSize,
      anchor, divisor, NPP_BORDER_REPLICATE, ctx));

    // Copy filtered image to the host
    checkCudaErrors(cudaMemcpyAsync(imgData, outMem,
      width * height, cudaMemcpyDeviceToHost, ctx.hStream));

    // Wait for stream to finish
    cudaStreamSynchronize(ctx.hStream);

    // Write data to new image file
    stbi_write_png("filtered_smiley.png", width, height, 1,
      imgData, step);

    // Free resources
    cudaFreeAsync(inMem, ctx.hStream);
    cudaFreeAsync(outMem, ctx.hStream);
    cudaStreamDestroy(ctx.hStream);
    stbi_image_free(imgData);
    return 0;
}
```

It's important to understand how the application defines the filter's kernel. The coefficients must be provided as 32-bit signed integers (`Npp32s`), and they need to be copied to a block in device memory. In this case, the kernel is represented by a 3-by-3 matrix, so `kernelSize` is set to (3, 3). The filter results don't need to be scaled in this example, so the `divisor` parameter is set to 1.

9.6 Summary

Processing images at high speed is one of CUDA's greatest strengths and NPPI provides several sublibraries for this purpose. This chapter has explored three of them: nppial for arithmetic and logic, nppig for geometry transformation, and nppif for image filtering.

The operations in nppial are easy to use and understand. They make it possible to perform simple operations on each pixel of an image, including addition, subtraction, multiplication, and division. To demonstrate their usage, an example project updates the contrast of an image by subtracting a midpoint from each image, scaling the result, and adding the midpoint again.

The functions in the nppig library make it possible to perform geometric transformations. The three types of transformations are mirroring (flipping), rotation, and resizing. The second example project of this chapter rotated an image by calling `nppiRotate_8u_C1R_Ctx` and then resized it by calling `nppiResizeSqrPixel_8u_C1R_Ctx`.

The last part of the chapter presented the topic of image filtering, which uses convolution to update pixels of an image. Convolution computes a two-dimensional dot product for each pixel of an image, taking the product of the pixel and its surrounding pixels with values of a kernel, and then replacing the pixel with the sum. This process makes it possible to add special effects, and the last example project demonstrates how the nppif library can be used to sharpen an image.

Chapter 10

Graphical Rendering with OpenGL

In this chapter, we'll concentrate on *graphics*. To be specific, the goal of this and the following chapters is to show how CUDA can be used to accelerate graphical rendering with OpenGL. The process of graphical rendering converts a model of three-dimensional figures and produces a two-dimensional array of pixels that can be drawn on a screen. Rendering executes in a loop, and as the properties of the figures change, the renderer updates the drawing.

To accelerate rendering, Nvidia makes it possible to create OpenGL models and update OpenGL textures using CUDA kernels. Before you can understand this, you need a solid understanding of OpenGL development. The goal of this chapter is to explain the fundamentals of OpenGL so that you can run simple applications. Later chapters will explain how to accelerate the rendering with CUDA.

The chapter starts by explaining how to install the libraries needed to build and execute the example projects. The next topic is the rendering pipeline, which is responsible for converting points in space to pixels in the screen. The following section explains how vertex data needs to be packaged for the OpenGL pipeline.

A large part of this chapter is devoted to coding OpenGL shaders. Like a CUDA kernel, a shader is a portion of code intended to run on the GPU. Unlike CUDA kernels, every shader has a specific purpose. At minimum, a vertex shader computes positions of the points in the scene and a fragment shader computes the colors associated with the scene's pixels.

The last part of the chapter explains how to use the FreeGLUT toolset, which makes it easy to display OpenGL graphics in a window. The example code presented in this chapter combines OpenGL and FreeGLUT to demonstrate how OpenGL rendering works. The next chapter will explain how the rendering can be accelerated with CUDA.

10.1 Installation

OpenGL renders graphics, but it doesn't provide any way to display the graphics. In the 1990s, Mark Kilgard created the OpenGL Utility Toolkit (GLUT) to make up for this shortcoming. GLUT is a simple, cross-platform toolset for creating windows capable of displaying OpenGL graphics. GLUT has fallen into disuse, but the open-source FreeGLUT is still developed and used.

The example projects in this and the next two chapters rely on OpenGL to render graphics and FreeGLUT to display the graphics. To build and run these projects, you'll need to install the header files and libraries for both toolsets.

Nvidia's driver provides an OpenGL library for your device, so you don't need to install a new OpenGL library. But Nvidia doesn't provide the header files needed to access the library in code.

10.1.1 Windows Installation

On Windows, you don't need to install a new OpenGL library. The main OpenGL library is OpenGL32.lib, and on my system, it's located in the C:\Program Files (x86)\Windows Kits\10\Lib\10.0.26100.0\um\x64 directory.

There are two main ways to obtain FreeGLUT libraries on Windows. The first involves visiting the FreeGLUT site at https://freeglut.sourceforge.net, downloading the source code, and building the libraries using CMake. This provides the most recent version (currently 3.6.0), but entails a fair amount of work.

The second way involves directly downloading pre-built FreeGLUT binaries. If you visit http://www.transmissionzero.co.uk/software/freeglut-devel and click one of the download links, you'll receive an archive containing the header file and two libraries needed to access FreeGLUT. This is convenient but unsafe, and provides version 3.0.0 of FreeGLUT, which is rather old.

The two FreeGLUT libraries are freeglut.lib and freeglut.dll. To build an application with FreeGLUT, it needs to access the freeglut.lib import library, which accesses the objects in freeglut.dll. You'll need to make sure that freeglut.dll is located where the loader can find it. One way to do this is to store the 64-bit version of the library in the C:\Windows\System32 folder.

Unfortunately, the OpenGL library on Windows only provides access to features from a very early version of OpenGL (version 1.1). To access modern features, you'll need to install a toolset like the OpenGL Extension Wrangler Library (GLEW). Installing GLEW requires four steps:

1. Visit the GLEW main site (https://glew.sourceforge.net) in a browser.
2. In the center of the page, click the link to download the GLEW binaries.
3. Unzip the archive when the download is finished.
4. Find the appropriate DLL in the bin folder and save it to an appropriate place on your system.

To build the example projects, the `GLUT_DIR` environment variable needs to be set to the directory containing FreeGLUT's libraries and header file. The `GLEW_DIR` environment variable needs to be set to the directory containing GLEW's libraries and header file.

10.1.2 Linux Installation

On Linux, the Nvidia driver provides an OpenGL library named libGL.so. But the header files need to be obtained through the open-source Mesa package. On Ubuntu and Debian systems, this can be obtained with the following command:

```
sudo apt install mesa-common-dev
```

If this executes successfully, the gl.h header file will be located in the /usr/include/GL folder.

This header doesn't provide all the functions we need, so it's necessary to install the OpenGL Extension Wrangler library, or GLEW. On Ubuntu and Debian systems, this can be obtained with the following command:

```
sudo apt install libglew-dev
```

To create the rendering window, the example code in this book relies on FreeGLUT. This is provided as an open-source package, and installing it on Linux is easy. On Ubuntu and Debian systems, the installation can be performed with the following command:

```
sudo apt install freeglut3-dev
```

In addition to installing OpenGL, GLEW, and GLUT, some Linux systems need to be directed to use the Nvidia GPU for graphics. On my system, I set the `__NV_PRIME_RENDER_OFFLOAD` environment variable to `1`. I also set the `__GLX_VENDOR_LIBRARY_NAME` environment variable to `"nvidia"`. The second variable tells OpenGL to rely on Nvidia's OpenGL driver for graphical rendering.

10.2 The Rendering Pipeline

There are many toolsets available for graphical rendering, and though the functions and libraries change, the rendering process remains similar. The goal is to convert points in three-dimensional space into pixels on a two-dimensional screen. The points in space are called *vertices*, and at minimum, applications need to specify where each vertex is located. Applications can also set the vertex's color, texture coordinates, and normal vector components. This per-vertex data is called *attribute data.*

Just as products are manufactured in stations along an assembly line, the steps needed to convert 3-D points to 2-D pixels form a *rendering pipeline*. In the early days of graphical programming, pipelines were *fixed*. In a fixed pipeline, the CPU sends vertex data to the GPU and has no further interaction with the rendering process.

Modern applications rely on *programmable* pipelines. In a programmable pipeline, certain stages can be customized with special code that executes on the GPU. Figure 10.1 illustrates a basic programmable pipeline, with the programmable stages shaded in dark gray.

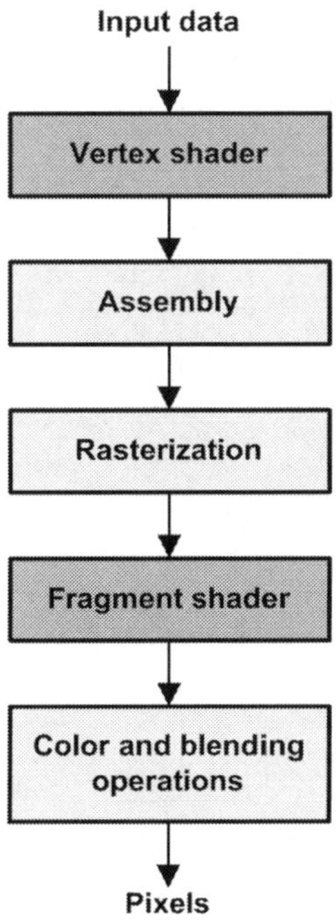

Figure 10.1 Programmable Pipeline

The programs executed in the programmable stages are called *shaders*. OpenGL supports several types of shaders, but this book focuses on vertex shaders and fragment shaders. These shaders must be written in the OpenGL Shading Language (GLSL).

After a vertex shader processes input data, the pipeline combines vertices into basic shapes called *primitives*, such as points, lines, or triangles. The process of creating primitives is called *assembly*. Later in this chapter, I'll explain how the assembly process can be customized in code.

After vertices have been assembled into primitives, the next step is *rasterization*, which breaks each primitive into a two-dimensional array of elements called *fragments*. Fragments are conceptually similar to pixels, but while a pixel only has a color and a location, a fragment has a color, a location, and a depth.

Once rasterization is finished, each fragment is processed by the fragment shader, which can customize the fragment's color and depth. Many fragment shaders use lighting algorithms to set the fragment's color depending on the sources of light in the model. One popular method for setting lighting in a scene is the Blinn-Phong lighting model.

When the last stage has completed, the result will be a two-dimensional array of pixels. These pixels are stored in a special memory region called a *framebuffer*, and OpenGL makes it possible to read and write directly to this buffer. However, framebuffer access lies outside the scope of this book.

10.3 Vertex Data

The first step of the rendering process is to define and package data that identifies the model's vertices. In OpenGL, working with vertex data requires at least four steps:

1. Place vertex data in special buffers called vertex buffer objects (VBOs).
2. Make vertex data accessible to the shaders by configuring vertex attribute objects (VAOs).
3. Compile shader code and load the shader executables onto the GPU.
4. Launch the rendering process.

This section will present each of these four steps and the functions needed to perform them. The coding process isn't simple, and it took me some time to understand the relationship between vertex buffers, vertex attributes, and shader variables. Once you start looking at example code and writing your own, these topics will become clearer.

10.3.1 Vertex Buffer Objects (VBOs)

As discussed in Chapter 3, CUDA makes it possible to allocate special blocks of device memory to store data for kernel processing. After allocating a block of device memory, an application initializes an array of data and then copies it to the device.

OpenGL relies on similar data structures to store vertex data, and they're called *vertex buffer objects*, or VBOs. Unfortunately, working with VBOs is significantly more complicated than working with CUDA device memory. There are four functions that need to be called, and Table 10.1 lists them in order.

Table 10.1

Vertex Buffer Object Functions

Function	Description
`glGenBuffers(GLsizei num, GLuint* descriptors)`	Create buffer objects and initialize an array of descriptors
`glBindBuffer(GLenum target, GLuint descriptor)`	Make the buffer object active
`glBufferData(GLenum target, GLsizeiptr size, const GLvoid* data, GLenum usage)`	Associate the buffer object with data
`void glDeleteBuffers(GLsizei num, const GLuint* descriptors)`	Delete buffer objects identified by the descriptors

The new data types may look strange, but they serve the same purposes as types in C and C++. A `GLuint` is an unsigned integer and a `GLenum` is an enumerated type.

In CUDA, blocks of device memory are accessed using regular pointers. In OpenGL, vertex buffer objects are accessed through unsigned integers called *descriptors*. When an application calls `glGenBuffers`, the first argument identifies the number of VBOs to create and the second argument points to the array of descriptors. The following code shows how this can be used to create six buffer objects:

```
GLuint vbos[6];
glGenBuffers(6, vbos);
```

One major difference between OpenGL and CUDA is the use of *targets*. An OpenGL target is an identifier that represents a specific purpose. When an application binds a data structure to a target, it's specifying the intended purpose of the data structure. Only one structure can be bound to a target at a time.

The `glBindBuffer` function binds one of the VBO descriptors to a target that identifies the nature of the data stored in the buffer. If a VBO is intended to store vertex attribute data, such as vertex coordinates and colors, the corresponding descriptor must be bound to the `GL_ARRAY_BUFFER` target. The following code binds the first descriptor (`vbos[0]`) to this target:

```
glBindBuffer(GL_ARRAY_BUFFER, vbo[0]);
```

After binding a VBO descriptor with a target, an application can associate the descriptor with data by calling `glBufferData`. The first argument sets the target, the second sets the size of the data, the third points to the data, and the last argument tells OpenGL how the data will be accessed.

For example, suppose all the vertex data is stored in an array named `vbo_data`, which contains 120 `float`s. The following code associates the data with the VBO represented by the descriptor `vbos[0]`:

```
glBufferData(GL_ARRAY_BUFFER, 120 * sizeof(GLfloat), vbo_data,
  GL_STATIC_DRAW);
```

The last argument, `GL_STATIC_DRAW`, tells OpenGL that the VBO's data will be set once and accessed frequently (with every draw operation). Table 10.2 lists the nine values that can be set for this argument.

Table 10.2

Buffer Usage Specifiers

Specifier	Description
`GL_STATIC_DRAW`	Data is meant for drawing, and will be set once and used frequently
`GL_STATIC_READ`	Data is meant to be read, and will be set once and used frequently
`GL_STATIC_COPY`	Data is meant to be read and written, and will be set once and used frequently
`GL_DYNAMIC_DRAW`	Data is meant for drawing, and will be modified repeatedly and used frequently
`GL_DYNAMIC_READ`	Data is meant to be read, and will be modified repeatedly and used frequently
`GL_DYNAMIC_COPY`	Data is meant to be read and written, and will be modified repeatedly and used frequently
`GL_STREAM_DRAW`	Data is meant for drawing, and will be modified once and used a few times

GL_STREAM_READ	Data is meant to be read, and will be modified once and used a few times
GL_STREAM_COPY	Data is meant to be read and written, and will be modified once and used a few times

When a VBO has been created, bound to a target, and associated with data, OpenGL will access its data during its draw operations. A later section will explain how these draw operations are performed in code.

After an application finishes using a VBO, it can unbind the VBO from the target by calling `glBindBuffer` with the descriptor set to 0. When the VBO is no longer needed, the application can deallocate its memory with `glDeleteBuffers`, which accepts the same parameters as `glGenBuffers`.

10.3.2 Vertex Attribute Objects (VAOs)

During the drawing process, OpenGL executes the vertex shader once for each vertex in the model, passing it input variables called *attributes*. To obtain these variables for the shader, OpenGL extracts data from the VBO currently bound to the `GL_ARRAY_BUFFER` target. Each input variable of the shader corresponds to a separate VBO, so if the shader accepts three variables, OpenGL will extract data from three VBOs.

Before OpenGL can extract data from a VBO, it needs to know which shader variable corresponds to the VBO and how the VBO's data is structured. To provide this information, an application needs to create vertex attribute objects, or VAOs.

Like VBOs, VAOs are identified by descriptors and only one VAO can be active at a time. Table 10.3 lists the functions available for working with VAOs.

Table 10.3

Vertex Attribute Object Functions

Function	Description
`glGenVertexArrays(GLsizei num, GLuint *descriptors)`	Create VAOs and initialize an array of descriptors
`glBindVertexArray(GLuint descriptor)`	Make the VAO active
`glEnableVertexAttribArray(GLuint index)`	Enable the vertex attribute array with the given index
`glDisableVertexAttribArray(GLuint index)`	Disable the vertex attribute array with the given index

`glVertexAttribPointer(GLuint index, GLint size, GLenum type, GLboolean norm, GLsizei stride, const GLvoid* pointer)`	Set the structure of the data to be extracted for a shader variable
`glDeleteVertexArrays(GLsizei n, const GLuint *arrays)`	Free memory for the given VAOs

The first two functions, `glGenVertexArrays` and `glBindVertexArray`, are similar to `glGenBuffers` and `glBindBuffer`. `glGenVertexArrays` creates one or more VAOs and initializes an array of descriptors. Once a VAO descriptor has been created, the VAO can be made active by calling `glBindVertexArray` with its descriptor. This is shown in the following code, which creates a single VAO and makes it active.

```
unsigned int vao;
glGenVertexArrays(1, &vao);
glBindVertexArray(vao);
```

Every variable received by the vertex shader is assigned a unique index. For example, vertex coordinates might be assigned an index of 0 and the vertex colors might be assigned an index of 1. When an application calls `glEnableVertexAttribArray` with an index, it tells OpenGL that data will be provided for the variable with that index.

The `glVertexAttribPointer` function identifies how the data inside a VBO should be extracted for a given variable. This accepts six arguments:

- `index` — the index of the vertex shader variable (vertex attribute) to receive the data
- `size` — number of values to be extracted
- `type` — the data type of each value, such as `GL_FLOAT` or `GL_UNSIGNED_INT`
- `norm` — whether integers should be converted to a normalized floating point range (between 0.0 and 1.0)
- `stride` — number of bytes between values to be extracted
- `pointer` — number of bytes following the start of the VBO to extract the first values

This may seem complicated, but it will become clear when you look at an example. A later section will walk through the code needed to set vertex coordinates and colors using `glVertexAttribPointer`.

Once the VBOs and VAOs have been created and configured, the data is ready to be delivered to shaders. But before a shader can process data, it needs to be compiled and loaded onto the GPU. This is the topic of the next section.

10.4 Compiling Shaders and Programs

If you look in the ch10_three_squares project, you'll see that it contains two text files named three_squares.vert and three_squares.frag. These files contain the code for the application's vertex shader and fragment shader. Before the application can draw anything, this code must be read from the files, compiled, and linked into a program.

To be precise, the process of working with OpenGL shaders consists of five steps:

1. Create shader objects and associate them with code.
2. Compile each shader object.
3. Create a program object and attach the shader objects.
4. Link the shader program to form a GPU executable.
5. Deploy the executable to the GPU.

Just as VBOs and VAOs are accessed using integer descriptors, shaders are accessed using integers called *shader objects*. Table 10.4 lists ten of the OpenGL functions that interact with shader objects.

Table 10.4

Shader Object Functions

Function	Description
`glCreateShader(GLenum shaderType)`	Creates a shader object with the given type
`glShaderSource(GLuint shader, GLsizei count, const GLchar** string, const GLint* length)`	Sets the source code for a shader object
`glCompileShader(GLuint shader)`	Compiles the source code for a shader object
`glGetShaderiv(GLuint shader, GLenum pname, GLint* params)`	Obtains information regarding a shader object
`glGetShaderInfoLog(GLuint shader, GLsizei maxLength, GLsizei* length, GLchar* infoLog)`	Returns the log containing information about a shader's compilation
`glCreateProgram()`	Creates an empty program object
`glAttachShader(GLuint program, GLuint shader)`	Attaches a shader object to a program object

glLinkProgram(GLuint program)	Create executable from program object
glUseProgram(GLuint program)	Makes the program object active
glDeleteProgram(GLuint program)	Deallocates program object

At minimum, every OpenGL application needs to have a vertex shader and a fragment shader, so each of the first four functions need to be called twice. The following code creates a shader object representing a vertex shader and a shader object representing the fragment shader.

```
GLuint vs, fs;

// Create vertex shader
vs = glCreateShader(GL_VERTEX_SHADER);

// Create fragment shader
fs = glCreateShader(GL_FRAGMENT_SHADER);
```

After the shader objects are created, the next step is to associate them with text (`char` arrays) containing source code. Suppose the source code for the vertex shader is in `vsCode` and the source code for the fragment shader is in `fsCode`. The application can associate the shader objects (`vs` and `fs`) with the code by calling `glShaderSource`. The following code shows how this works:

```
glShaderSource(vs, 1, (const char**)&vsCode, &vsLength);
glShaderSource(fs, 1, (const char**)&fsCode, &fsLength);
```

In this code, `vsLength` is the number of characters in the `vsCode` array and `fsLength` is the number of characters in the `fsCode` array. Once the source code has been associated with a shader object, the shader can be compiled by calling `glCompileShader` with the object. To demonstrate, the following code compiles the vertex shader and fragment shader.

```
glCompileShader(vs);
glCompileShader(fs);
```

The `glCompileShader` function returns `void`, and doesn't provide any notification if the compilation fails. To get information, applications need to call a function called `glGetShaderiv`, which accepts a parameter and provides the corresponding data through a pointer. If the parameter is set to `GL_COMPILE_STATUS`, the function will return `GL_TRUE` if the compilation succeeded and `GL_FALSE` if it failed.

After the shaders have been compiled, the application needs to link them into a new type of object called a *program*. Program objects are created with the `glCreateProgram` function, and the following code shows how it works:

```
Gluint prog = glCreateProgram();
```

Next, the application's shader objects must be associated with the program using `glAttachShader`. The following code attaches the vertex shader (`vs`) and the fragment shader (`fs`) to the program object `prog`.

```
glAttachShader(prog, vs);
glAttachShader(prog, fs);
```

Once the attributes are associated with names and the shaders are attached to the program, the program can be linked and installed with the following code:

```
glLinkProgram(prog);
glUseProgram(prog);
```

After these functions execute, the program and its attached shaders will be deployed to the GPU. The GPU will execute them when it comes time to render graphics. In OpenGL parlance, the GPU's rendering is referred to as *drawing*, and the next section explains how applications tell the GPU to perform draw operations.

10.5 Draw Operations

After an application defines its vertex data and deploys shaders to the GPU, the only task remaining is to launch the rendering process. OpenGL provides several functions for this purpose, and the simplest is `glDrawArrays`, whose signature is given as follows:

```
void glDrawArrays(GLenum mode, GLint startIndex,
  GLsizei numIndices);
```

The first parameter tells the renderer what shapes to form from the model's vertices. These shapes are called *primitives*, and Table 10.5 lists seven of the different identifiers that can be set.

Table 10.5
Primitive Identifiers

Identifier	Description
`GL_POINTS`	Series of individual points
`GL_LINES`	Each pair of vertices form a separate line
`GL_LINE_STRIP`	Each vertex connected to its successive vertex
`GL_LINE_LOOP`	Each vertex connected to its successive vertex, last vertex connected back to first
`GL_TRIANGLES`	Each set of three vertices form a separate triangle
`GL_TRIANGLE_STRIP`	Forms triangle with first three vertices, each succeeding vertex forms a triangle with the two preceding it
`GL_TRIANGLE_FAN`	Forms triangle with first three vertices, each succeeding two vertices form triangle with first vertex

These identifiers tell OpenGL how to connect the dots (vertices) whose attributes are stored within the active VAO. Figure 10.2 depicts six of the different drawing modes.

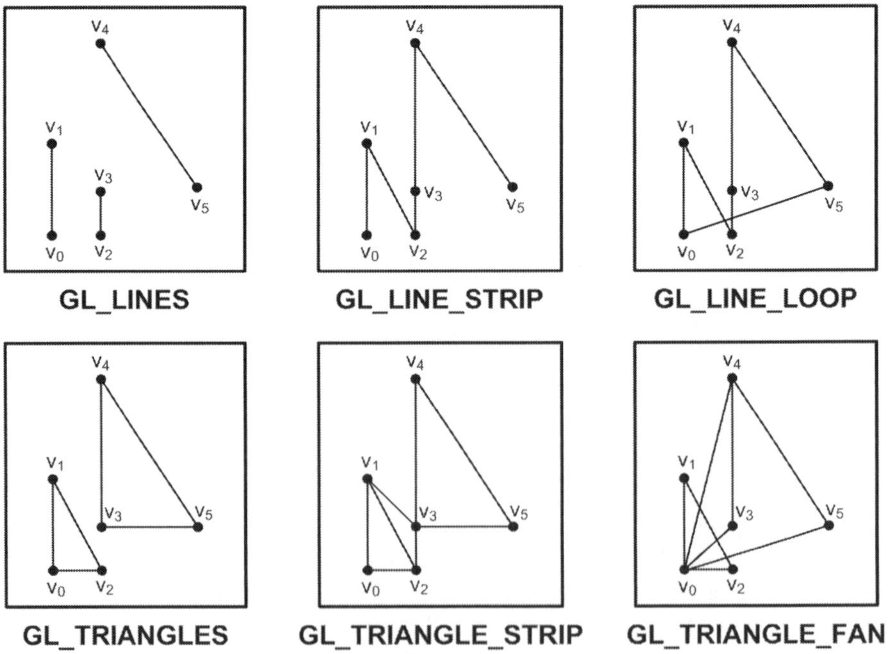

Figure 10.2 OpenGL Drawing Modes and Primitives

The simplest drawing modes are `GL_POINTS`, `GL_LINES`, and `GL_TRIANGLES`. When these modes are set, each primitive is drawn separately, without connections to others. That is, when `GL_LINES` is set, each pair of vertices is drawn as a separate line. When `GL_TRIANGLES` is set, each triple of vertices forms a separate triangle.

The `GL_LINE_STRIP` and `GL_TRIANGLE_STRIP` modes create interconnected primitives from each successive vertex. If `GL_LINE_STRIP` is set, the renderer simply draws a line from the first vertex to the second, third, fourth, and so on. If `GL_TRIANGLE_STRIP` is set, then each successive vertex forms a triangle with the two vertices preceding it. For example, if there are four vertices, v_0 to v_3, this mode will cause two triangles to be drawn: v_0-v_1-v_2 and v_1-v_2-v_3.

Once the drawing mode is set, the second and third arguments of `glDrawArrays` specify which vertices will be drawn. The second parameter identifies the starting index and the third parameter specifies the number of vertices. For example, the following code tells the renderer to draw six vertices as separate triangles:

```
glDrawArrays(GL_TRIANGLES, 0, 5);
```

Keep in mind that the draw operation only accesses data associated with the active VAO. An application with multiple VAOs will need to make each VAO active before calling `glDrawArrays`.

10.6 Shader Development

OpenGL shaders are coded with the OpenGL Shading Language (GLSL). If you understand how to write CUDA kernels, coding with GLSL won't present significant difficulty.

This section provides an overview of OpenGL shader development. It starts by presenting the data types and functions that shaders can access. Then it explores the two types of shader types required by all modern OpenGL applications: vertex shaders and fragment shaders.

10.6.1 Overview of Shader Coding

In my opinion, the best way for a CUDA developer to understand GLSL shaders is to see how they compare with CUDA kernels. Table 10.6 presents a list of similarities and differences between the two.

Table 10.6

Shaders and Kernels: Similarities and Differences

Similarities	Differences
Shaders and kernels are implemented as C-like functions that return void	Shader functions must be called main. Kernels can take any name, but must be preceded by the __global modifier.
Shaders and kernels process data in vectors and provide operators for numbers and vectors	Shaders support matrix data types. Basic CUDA kernels don't.
Shaders and kernels are math-oriented: no strings and no access to stdio.h or similar routines from the C Standard Library	Shader variables are accessed by index, which can be set with the layout parameter. Kernel variables are provided as function parameters.
Shaders and kernels can execute functions that closely resemble those in traditional math.h	Different variable modifiers that identify storage location and read/write ability

The fundamental difference between kernels and shaders is that, while a kernel can serve multiple purposes, each shader in an OpenGL application has a specific role to play. At the very least, an OpenGL application must have one shader to process vertices and one to process fragments.

Each shader needs to have a function called `main`. The following code presents a simple shader that sets an output vector variable, `out_vec`, equal to twice the value of an input vector variable, `in_vec`:

```
#version 330 core

layout (location = 0) in vec4 in_vec;
out vec4 out_vec;

void main(void) {
   out_vec = in_vec * 2;
}
```

As shown, the shader starts by specifying its version with a line that begins with the `#version` directive. The latest version of GLSL is 4.60, but to ensure maximum compatibility, it's safe to set the version to 3.30.

The `main` function doesn't accept or return any parameters, but operates on variables declared outside the function. This shader has two external variables named `in_vec` and `out_vec`. `in_vec` is preceded by the `in` keyword, which means the shader receives the variable as input. `out_vec` is preceded by the `out` keyword, which means the shader produces the variable as output.

The `in_vec` variable is also preceded by a `layout` specifier that sets its `location` to `0`. This provides the variable with an index, which means the OpenGL code running on the host can access the variable through the index of 0.

10.6.2 Data Types

While CUDA kernels support operations involving scalars and vectors, shaders support operations on scalars, vectors, and matrices. Table 10.7 presents an abridged set of the data types available:

Table 10.7

GLSL Data Types (Not Including Samplers)

Data Type	Description
bool/int/uint/float/double	Scalars: boolean, integer, unsigned integer, float, and double
bvec2/bvec3/bvec4	Vectors containing boolean values
ivec2/ivec3/ivec4	Vectors containing signed integers
uvec2/uvec3/uvec4	Vectors containing unsigned integers
vec2/vec3/vec4	Vectors containing float values
mat2/mat3/mat4	Square matrices containing float values
mat2x3/mat2x4/mat3x2/ mat3x4/mat4x2/mat4x3	Rectangular matrices containing float values
dmat2/dmat3/dmat4	Square matrices containing double values
dmat2x3/dmat2x4/dmat3x2/ dmat3x4/dmat4x2/dmat4x3	Rectangular matrices containing double values

GLSL's vector types have different names than CUDA's vector types, and they're initialized by surrounding elements in parentheses. For example, the following code initializes a vector whose four components equal 3.0.

```
vec4 v = (vec4)(3.0, 3.0, 3.0, 3.0);
```

Shaders can access the components of a vector using the dot-suffix method:

- `x, y, z, w` — identify coordinate elements or vector components
- `r, g, b, a` — identify channels of a pixel's color
- `s, t, p, q` — identify elements of texture coordinates

For example, if `vec1` and `vec2` both contain four elements, a shader can set `vec2` equal to the reverse of `vec1` with either of the following lines of code:

```
vec2 = vec1.wzyx
vec2 = vec1.abgr
vec2 = vec1.qpts
```

Working with matrices is similar to coding with vectors, but there's one important point to keep in mind: the elements need to be given in *column-major* order. This means that elements must be given one column at a time. For example, suppose a shader wants to create an `mat4` called `m` whose rows are (1.0, 2.0, 3.0, 4.0), (5.0, 6.0, 7.0, 8.0), (9.0, 10.0, 11.0, 12.0), and (13.0, 14.0, 15.0, 16.0). It would use the following code:

```
mat4 m = (mat4)(1.0, 5.0, 9.0, 13.0, 2.0, 6.0, 10.0, 14.0, 3.0,
  7.0, 11.0, 15.0, 4.0, 8.0, 12.0, 16.0);
```

In addition to providing data types for scalars, vectors, and matrices, GLSL supports simple math operations using traditional operators. It also provides several functions for comparison, exponentiation, trigonometry, and geometry.

10.6.3 Vertex Shaders

The first type of shader required in an OpenGL application is the vertex shader. This runs on the GPU and executes once for every vertex in the model. It receives attribute data from the host application, which can include coordinates, colors, texture coordinates, and normal vectors.

Vertex shaders can't add or remove vertices from the model, but it can change the vertices' attributes, such as their coordinates or colors. It may seem odd to change these attributes immediately after the host initialized them, but there are good reasons for this. If you want the renderer to present the model from a given angle, the vertex shader can update the vertices' coordinates. Similarly, to change the way an object in the model reflects light, the vertex shader can change the vertices' normal vectors.

A vertex shader must set the position of each vertex it processes. More precisely, every vertex shader needs to provide a value for a built-in variable called `gl_Position`. This is a `vec4`, so the shader needs to provide four components for each position.

An example will clarify how this works. Suppose the host application creates an attribute called `in_location` that identifies vertex coordinates within a VBO. In this case, a simple vertex shader would look like the following:

```
layout (location = 0) in vec4 in_location;

void main(void) {
   gl_Position = in_location;
}
```

This shader sets the position of each vertex equal to the `in_location` attribute provided by the host. This may seem trivial, but without this, the renderer won't know how to set the vertex's position.

Most vertex shaders perform more work than this. In many cases, vertex shaders use matrix-vector multiplication to update vertices' positions. As explained in Chapter 6, the multiplication of an n-element vector by an n-by-n matrix produces a second n-element vector. If coded correctly, a matrix can be used to rotate, move, or scale vertices in a model. These operations are collectively called *transformations* and their corresponding matrices are called transformation matrices.

For example, the ch10_three_squares project creates a model and rotates it so that the squares can be viewed from a specific viewpoint. This rotation is accomplished by the vertex shader, whose code is contained in the ch10_three_squares/three_squares.vert source file. Listing 10.1 shows what this code looks like.

Listing 10.1 ch10_three_squares/three_squares.vert

```
#version 330 core
layout (location = 0) in vec3 in_coords;
layout (location = 1) in vec3 in_color;
out vec3 new_color;

void main(void) {

  // Set the vertex color
  new_color = in_color;

  // Create a rotation matrix
  mat3x3 rot_matrix = mat3x3(0.707, 0.641, -0.299,
                             -0.707, 0.641, -0.299,
                             -0.000, 0.423,  0.906);

  // Rotate the vertex
  vec3 coords = rot_matrix * in_coords;

  // Set the vertex position
  gl_Position = vec4(coords, 1.0);
}
```

As shown, the vertex shader receives two attributes for each vertex. The `in_coords` attribute identifies the initial position of the vertex and `in_color` identifies the color of the vertex. The shader creates a 3-by-3 matrix called `rot_matrix` and multiplies it by the vertex's coordinates, thereby transforming the vertex's position.

The host application associates the color attribute with the name `in_color`. The vertex shader receives this attribute but doesn't modify its value. Instead, it creates another `vec3` variable called `new_color` and passes this to the next stage. In this simple application, the next stage is the fragment shader, which is the topic of the next section.

10.6.4 Fragment Shaders

After the vertex shader completes its processing, vertices will be combined into primitives such as points, lines, or polygons. Then OpenGL will convert these three-dimensional figures into two-dimensional shapes composed of pixels. This conversion process is called *rasterization*.

But the pixels produced by rasterization aren't necessarily the final pixels displayed by the renderer. Instead, OpenGL refers to the preliminary pixels as *fragments*. The primary goal of a fragment shader is to accept the fragment colors produced by rasterization and set the colors of the pixels in the final display.

Just as the vertex shader executes once for each vertex in the model, the fragment shader executes once for each fragment. The shader can't change a fragment's position or access other fragments, but it can serve the following five functions:

- Apply texture data
- Set the fragment's depth
- Configure lighting
- Bump-mapping
- Additional effects such as fog and shadow

To properly render a three-dimensional object like a sphere, the sphere needs to reflect light differently from different points on its surface. To make this possible, researchers have devised lighting models that determine how the colors at each point need to be changed. The Blinn-Phong lighting model is popular, and computes ambient, diffuse, and specular components of light.

In the ch10_three_squares project, the only purpose of the fragment shader is to set the fragment's output color. This is easy because no built-in variables are needed. If the fragment shader only has one output variable, the fragment's color will be set to this value. This is shown in Listing 10.2.

Listing 10.2 ch10_three_squares/three_squares.frag

```
#version 330 core

in  vec3 new_color;
out vec4 out_color;

void main(void) {

  // Update the color
  vec3 tmp_color = new_color + vec3(0.25f, 0.25f, 0.25f);

  // Set the fragment's color
  out_color = vec4(tmp_color, 1.0);
}
```

In this listing, `out_color` is the only variable declared with the `out` modifier, so the renderer will set the fragment's color to its value. The input variable, `new_color`, corresponds to the output variable (`new_color`) of the vertex shader. As shown in listing 10.1, this variable contains the color attribute defined by the host application.

The vertex shader passes `new_color` to the fragment shader, but the value of the color may not be the same as the color attribute defined on the host. This is because pixel colors may be interpolated before reaching the fragment shader. The fragment shader can't perform interpolation itself because it can't access colors of other fragments.

10.7 Creating Windows with GLUT

Figuratively speaking, you could say that OpenGL paints the picture and GLUT provides the frame and canvas. Technically, GLUT creates the window that OpenGL uses to render graphics. GLUT isn't the only toolset that makes this possible, and both Qt and GTK allow you to create full-featured GUIs with OpenGL. But in my opinion, GLUT is the easiest to use. This section presents GLUT's functions and shows how they can be used to form canvases for OpenGL.

10.7.1 Creating the Window

In any GLUT-based application, the first steps involve loading the GLUT runtime, configuring window properties, and creating a data structure to represent the window. Table 10.8 lists the six functions that initialize GLUT windows.

Table 10.8

GLUT Initialization Functions

Function	Description
glutInit(int* argc, char** argv)	Launches the GLUT runtime
glutInitWindowSize(int w, int h)	Sets the window's width and height
glutInitWindowPosition(int x, int y)	Sets the location of the window's initial position
glutFullScreen()	Configures the window to occupy the entire screen
glutInitDisplayMode(uint mode)	Sets the window's display mode, which identifies pixel format and color buffering
glutCreateWindow(char* title)	Creates a window with the given title, returns a numeric descriptor for the window

Of these functions, `glutInit` is the most important and must be called first. Once this is done, `glutInitWindowSize` sets the size of the desired window and `glutInitWindowPosition` sets its position in the screen. The `glutCreateWindow` function should be invoked after all other GLUT configuration functions have been called.

The `glutInitDisplayMode` function configures properties related to the window's pixel content and display characteristics. It accepts a single parameter that can be set to one of the constants in Table 10.9 or an OR-ed combination thereof.

Table 10.9

GLUT Display Modes

Data Type	Description
GLUT_RGB/GLUT_RGBA/GLUT_INDEX/ GLUT_LUMINANCE	Sets the channel format used for pixels within the window
GLUT_SINGLE/GLUT_DOUBLE	Sets whether the window uses single-buffering or double-buffering
GLUT_ACCUM/GLUT_ALPHA/GLUT_DEPTH/ GLUT_STENCIL	Identifies buffers to store data related to OpenGL buffering
GLUT_MULTISAMPLE	Enable full-screen anti-aliasing

The GLUT_RGB and GLUT_SINGLE modes are enabled by default. But throughout this book, the example code sets the display mode to GLUT_RGB | GLUT_DOUBLE under the assumption that modern hardware can handle double-buffered graphics. As an example, the following code creates a 100-by-100 pixel window at the location (0, 0).

```
glutInit(&argc, argv);
glutInitWindowSize(500, 150);
glutInitWindowPosition(0, 0);
glutInitDisplayMode(GLUT_RGB | GLUT_DOUBLE);
int window = glutCreateWindow("Title");
```

This code creates the data structure that GLUT uses to represent the window, but it doesn't actually display a new window. This requires event handling.

10.7.2 Event Handling

Event-handling is a major concern when coding graphical user interfaces (GUIs), which need to respond to mouseovers, mouse clicks, keystrokes, and resizing events. GLUT makes this simple by providing event-handling functions that accept pointers to functions called *callback functions*, which respond to the corresponding event. Table 10.10 lists a subset of GLUT's event-handling functions and the events they respond to.

Table 10.10

GLUT Event Handling Functions

Function	Event
`glutDisplayFunc(void(*func)(void))`	Window display
`glutMouseFunc(void(*func) (int button, int state, int x, int y))`	Mouse click/release at the given coordinates
`glutReshapeFunc(void(*func) (int width, int height))`	Resizing of the GLUT window
`glutKeyboardFunc(void(*func) (unsigned char key, int x, int y))`	Keystroke
`glutVisibilityFunc(void(*func) (int state))`	Change in the window's visibility status
`glutTimerFunc(unsigned int ms_time, void(*func)(int value), int value)`	Passage of the specified amount of time
`glutIdleFunc(void (*func)(void))`	Called when events aren't being received

The callback function identified by `glutDisplayFunc` tells the window how to display itself. For this reason, it's the only callback registration function that *must* be implemented in a GLUT application. To see how it works, it helps to look at an example. Suppose you have a function called `display` given as follows:

```
void display() {
    ...
}
```

The following code ensures that GLUT will call this function when it needs to redraw the pixels in the window:

```
glutDisplayFunc(&display);
```

As a further example, the following code creates a function that receives reshaping events and then registers it as a GLUT callback function:

```
void reshape(int width, int height) {
    ...
}

glutReshapeFunc(&reshape);
```

In both of these examples, the name of the callback function is taken from the name of the corresponding GLUT registration function. This convention makes coding easier and is employed throughout this book.

10.7.3 Displaying the Window

After the window has been initialized and the event-handling routines have been registered, the next step is to display the window. The function that makes this possible is `glutMainLoop`, which doesn't accept any parameters or return any values. It starts the GLUT processing cycle that receives events and invokes their corresponding callback functions as needed. The loop continues until the application terminates.

The code in Listing 10.3 provides a simple example of how GLUT can be used. It calls `glutInit`, sets the window's size to (500, 150), its location to (200, 100), and then calls `glutCreateWindow`. It also registers two callback functions: one that executes when the window draws its pixels and one that responds to resizing events.

Listing 10.3 Simple GLUT Example

```
#include <GL/freeglut.h>
#include <stdio.h>

// Called when the window is reshaped
void reshape(int width, int height) {
  printf("New dimensions: %d %d\n", width, height);
}

// Called to display the window
void display() {
  glClear(GL_COLOR_BUFFER_BIT);
  printf("Displaying the window\n");
  glutSwapBuffers();
}

int main(int argc, char** argv) {

  // Initializes and configures the window
  glutInit(&argc, argv);
  glutInitDisplayMode(GLUT_RGB | GLUT_DOUBLE);
  glutInitWindowSize(500, 150);
  glutInitWindowPosition(200, 100);

  // Creates the window
  glutCreateWindow("Introducing GLUT");

  // Event handling
  glutDisplayFunc(display);
  glutReshapeFunc(reshape);

  // Sets the initial color
  glClearColor(1.0, 1.0, 1.0, 0.0);

  // Launches the window
  glutMainLoop();
  return 0;
}
```

The callback functions that handle display and resizing events become particularly important for windows that display OpenGL rendering. As OpenGL-specific code is added, the content of the window becomes much more interesting.

10.8 Combining OpenGL and GLUT

To make a GLUT window capable of displaying OpenGL graphics, three steps are necessary:

1. During the window's initialization, define vertex data and vertex objects, and compile shaders into a program.
2. Set GLUT's `reshape` callback to invoke `glViewport`, which tells OpenGL how large the display canvas is.
3. Set GLUT's `display` callback to invoke `glDrawArrays`, which will render the 3-D model in the GLUT window.

These steps are straightforward so long as you remember when the GLUT event handling callbacks are invoked. For example, the reshape callback is called whenever the window is resized. The display callback is called whenever the window needs to draw itself.

10.8.1 Initializing GLUT and OpenGL

If you look at the main.cpp file in the ch10_ three_squares project, you'll see that the `main` function calls two new functions:

- `initBuffers` — initializes the vertex data, vertex array objects, and vertex buffer objects
- `initShaders` — reads shader code from the shader files, compiles the shader code, and links the code into a program

These two functions handle the OpenGL initialization tasks for the application. This discussion explains what both functions accomplish.

Initializing Buffers

The goal of `initBuffers` is to create the vertex array objects (VAOs) and vertex buffer objects (VBOs) needed to store the application's vertex data. This data consists of six arrays: three arrays containing vertex coordinates and three arrays containing vertex colors. Each array contains data for four vertices, which makes sense because our goal is to render the three squares depicted in Figure 10.3 (on the following page).

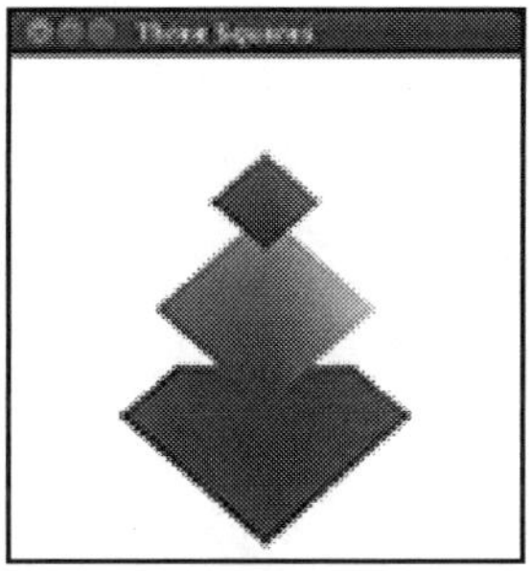

Figure 10.3 The Three Squares Rendering

The application draws three distinct objects, so `initBuffers` creates three vertex array objects (VAOs). Each object needs two pieces of information: its color and the coordinates of its vertices. Each piece of information is stored in a separate VBO, so the application creates six VBOs:

```
GLuint vao[3], vbo[6];
glGenVertexArrays(3, vao);
glGenBuffers(6, vbo);
```

Now let's take a close look at the vertex data for the topmost square. The application needs to set the location and color of each vertex, and Figure 10.4 gives an idea of how these attributes are set.

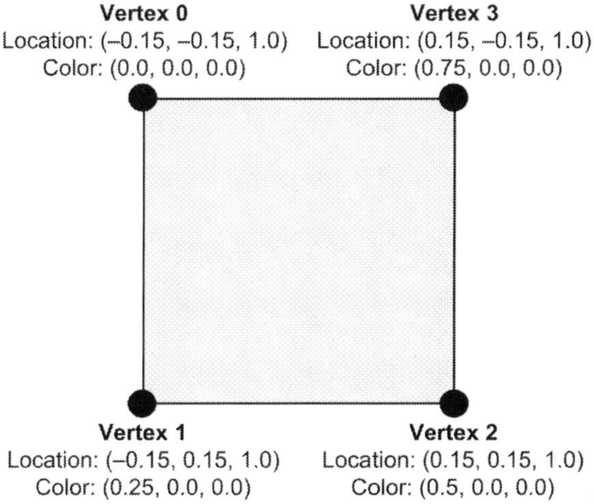

Figure 10.4 Vertex Data for the First Square

In code, the coordinates and colors of the first square's vertices are provided with the following arrays:

```
// The coordinates of the vertices of the first square
GLfloat firstCoords[] = { -0.15f, -0.15f, 1.0f,
                          -0.15f,  0.15f, 1.0f,
                           0.15f,  0.15f, 1.0f,
                           0.15f, -0.15f, 1.0f };

// The colors of the vertices of the first square
GLfloat firstColors[] = { 0.0f,  0.0f, 0.0f,
                          0.25f, 0.0f, 0.0f,
                          0.50f, 0.0f, 0.0f,
                          0.75f, 0.0f, 0.0f };
```

Looking at this code, there are three points to keep in mind:

- The x-coordinates left of the center are negative and the x-coordinates right of the center are positive. The y-coordinates above the center are negative and the y-coordinates below the center are positive.
- The RGB components of the vertex colors are given as normalized floating-point values, running from 0.0 to 1.0.
- The vertex coordinates will be updated in the vertex shader. The vertex colors will be updated in the fragment shader.

Before OpenGL can access this data, the application needs to associate these arrays with vertex buffer objects. The following code associates `firstCoords` with `vbo[0]` and `firstColors` with `vbo[1]`.

```
// VBO for coordinates of first square
glBindVertexArray(vao[0]);
glBindBuffer(GL_ARRAY_BUFFER, vbo[0]);
glBufferData(GL_ARRAY_BUFFER, 12 * sizeof(GLfloat),
  firstCoords, GL_STATIC_DRAW);
glVertexAttribPointer(0, 3, GL_FLOAT, GL_FALSE, 0, 0);
glEnableVertexAttribArray(0);

// VBO for colors of first square
glBindBuffer(GL_ARRAY_BUFFER, vbo[1]);
glBufferData(GL_ARRAY_BUFFER, 12 * sizeof(GLfloat),
  firstColors, GL_STATIC_DRAW);
glVertexAttribPointer(1, 3, GL_FLOAT, GL_FALSE, 0, 0);
glEnableVertexAttribArray(1);
```

Before the VBOs are bound and configured, glBindVertexArray makes the first vertex array object (VAO) active. As a result, all configured vertex data will be associated with this VAO.

Once the VAO is set, the first VBO is bound to the GL_ARRAY_BUFFER target. Then glBufferData associates the firstCoords array with the VBO. The array contains twelve floats (three coordinates per vertex), so the second argument is set to 12 * sizeof(GLfloat). glVertexAttribPointer tells OpenGL that three values should be accessed at a time, and that the data should be provided to the shader variable with the index of 0.

Once the first VBO is configured, the next lines configure the second VBO. Now glBufferData associates the firstColors array with the VBO. glVertexAttribPointer tells OpenGL that the data should be provided to the shader variable with the index of 1.

Once these VBOs are configured, initBuffers binds and configures the next two VAOs. Each of these contains similar vertex data, and use essentially the same code as that shown above.

Initializing Shaders

After configuring the vertex data, the main function calls initShaders, which starts by creating one shader object for the vertex shader and one for the fragment shader.

```
GLuint vs, fs;
vs = glCreateShader(GL_VERTEX_SHADER);
fs = glCreateShader(GL_FRAGMENT_SHADER);
```

The code for the vertex shader is in three_squares.vert and the code for the fragment shader is in three_squares.frag. To read these files, initShaders calls the readFile function with a macro identifying the file name.

```
char *vsSource, *fsSource;
vs_source = readFile(VERTEX_SHADER, &vsLength);
fs_source = readFile(FRAGMENT_SHADER, &fsLength);
```

After the source code has been read for each shader, initShaders calls glShaderSource to associate the source code with the corresponding shader object.

```
glShaderSource(vs, 1, (const char**)&vsSource, &vsLength);
glShaderSource(fs, 1, (const char**)&fsSource, &fsLength);
```

Now that the shader objects have associated source code, `initShaders` calls `compileShader` with each shader object. The code for the `compileShader` function is given as follows:

```
glCompileShader(shader);
glGetShaderiv(shader, GL_COMPILE_STATUS, &success);
if (!success) {
    glGetShaderiv(shader, GL_INFO_LOG_LENGTH, &log_size);
    log = (char*)malloc(log_size + 1);
    log[log_size] = '\0';
    glGetShaderInfoLog(shader, log_size + 1, NULL, log);
    printf("%s\n", log);
    free(log);
    exit(1);
}
```

This calls `glCompileShader` to perform the compilation and `glGetShaderiv` to check if it succeeded or failed. If compilation failed, the function calls `glGetShaderiv` to get the log size and `glGetShaderInfoLog` to retrieve the log. Then the content of the log is printed and the application exits.

If the shaders were compiled successfully, `initShaders` calls `glCreateProgram` to create a program object.

```
GLuint prog;
prog = glCreateProgram();
```

After the program object is created, `glAttachShader` is needed to associate each shader object with the program.

```
glAttachShader(prog, vs);
glAttachShader(prog, fs);
```

Once the shaders are attached to the program, `initShaders` calls `glLinkProgram` to link the program object and `glUseProgram` to associate the program object with the current rendering state.

```
glLinkProgram(prog);
glUseProgram(prog);
```

Now that these functions are called, the shaders will be invoked as part of the rendering pipeline. There's no need to call `glDeleteProgram` because the program will be deallocated at the application's conclusion.

10.8.2 Setting the Viewport

Before a scene can be rendered, OpenGL needs to know the size and location of the region to serve as the rendering canvas. This region is called a *viewport* and it's defined with the `glViewport` function, whose signature is given as follows:

```
glViewport(GLint x, GLint y, GLsizei width, GLsizei height)
```

The operation of this function is straightforward. The `x` and `y` parameters identify the upper-left position of the region, and `width` and height `identify` the size. If a user moves or resizes the GLUT window, the viewport needs to be updated. Therefore, the `resize` callback function calls `glViewport` in the following manner:

```
void reshape(int w, int h) {
   glViewport(0, 0, (GLsizei)w, (GLsizei)h);
}
```

Keep in mind that the `x` and `y` parameters (0 and 0 in this example) identify the region's location relative to the GLUT window's client area, not the absolute location within the screen.

10.8.3 Rendering the Model

The GLUT window calls the `display` callback function every time it needs to redraw its content. The OpenGL function `glDrawArrays` launches the rendering process, so it makes sense that the `display` callback should invoke `glDrawArrays`. But first, the callback calls `glClear` to reset the rendering area.

```
glClear(GL_COLOR_BUFFER_BIT | GL_DEPTH_BUFFER_BIT);
```

The application draws the third square first, then the second square, and then the first square. This ensures that upper squares overlap squares below them.

```
// Draw the third square
glBindVertexArray(vao[2]);
glDrawArrays(GL_TRIANGLE_FAN, 0, 4);

// Draw the second square
glBindVertexArray(vao[1]);
glDrawArrays(GL_TRIANGLE_FAN, 0, 4);
```

```
// Draw the first square
glBindVertexArray(vao[0]);
glDrawArrays(GL_TRIANGLE_FAN, 0, 4);
```

In each case, the `display` function calls `glBindVertexArray` to retrieve the data associated with that VAO. Then it calls `glDrawArrays` to draw the square. There is no primitive for squares, so the first argument is set to `GL_TRIANGLE_FAN` to tell OpenGL to draw two connected triangles. The second argument is set to `0` to specify that the first vertex to be drawn is the first vertex in the VBO. The last argument is set to `4` because four vertices need to be drawn.

After the squares are drawn, the display function unbinds all VAOs by calling `glBindVertexArray` with an argument of `0`.

```
glBindVertexArray(0);
```

So far, the draw operations have updated the *framebuffer*, which is the block of memory dedicated to storing pixels to be displayed. The last step of the `display` function tells GLUT that the content of the framebuffer should be used to set the pixels in the current window.

```
glutSwapBuffers();
```

The pixels from the framebuffer will remain displayed in the window until the `display` callback is invoked again.

10.9 Adding Texture

Many applications need to cover the surfaces of a model with pixels taken from images. OpenGL makes this possible through *textures*. If you think of vertices as forming a skeleton, then textures form the skin. On the following page, Figure 10.5 shows what's happening. The ch10_texture_squares project creates textures from the three images on the left and uses them to cover the three squares.

Conceptually, working with textures is like putting up wallpaper. In both cases, blank surfaces are made more appealing through decorative images. The corners of the pattern need to be carefully matched to the corners of the surface. It's also important to make sure that the pattern is oriented correctly as it's applied.

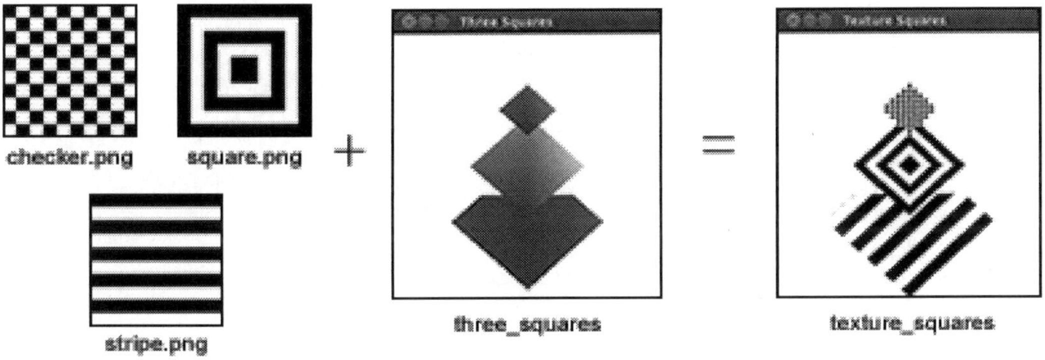

Figure 10.5 Using Texture in the ch10_texture_squares Project

But textures are easier to work with than wallpaper because there's no need to do any measuring. OpenGL will stretch or shrink the texture as needed to fit the space. Also, if a texture needs to be minimized, you can tell OpenGL to use miniaturized versions of the texture called *mipmaps*.

The goal of this discussion is to explain how to convert image data into textures and apply the textures to surfaces in a model. There are three steps:

1. In the host application, create a texture object and configure it with data from an image file.
2. In the vertex shader, match the texture's coordinates to vertices of the model.
3. In the fragment shader, set the output color equal to the texture color.

After discussing the functions needed to implement these steps, we'll look at the ch10_texture_squares project.

10.9.1 Creating Textures

Like vertex buffer objects, textures in OpenGL are accessed through unsigned integers called descriptors. The process of working with textures consists of four steps:

1. Create a descriptor for each texture.
2. Make a texture active.
3. Set characteristics of the active texture.
4. Associate the active texture with data.

The functions that create, configure, and delete textures are also similar to those used for VBOs. Table 10.11 lists six of the most important functions related to textures.

Table 10.11

Texture Functions

Function	Event
`glGenTextures(GLsizei num, GLuint* descriptors)`	Create texture objects and initialize an array of descriptors
`glBindTexture(GLenum target, GLuint descriptor)`	Make the texture object active for a given target
`glPixelStorei(GLenum param_name, GLint param)`	Identify characteristics about how the pixel data is stored
`glTexParameteri(GLenum target, GLenum param_name, GLint param)`	Identify characteristics about how the texture should be used
`glTexImage2D(GLenum target, GLint level, GLint intformat, GLsizei width, GLsizei height, GLint border, GLenum format, GLenum type, const GLvoid* data)`	Associates image data with the texture
`glDeleteTextures(GLsizei num, const GLuint* descriptors)`	Delete texture objects identified by the descriptors

The first two functions, `glGenTextures` and `glBindTexture`, serve the same functions as the `glGenBuffers` and `glBindBuffer` functions discussed earlier. The first creates one or more texture objects and the second makes a texture active by binding it to a target. For `glBindTexture`, the available targets are `GL_TEXTURE_1D`, `GL_TEXTURE_2D`, `GL_TEXTURE_3D`, and `GL_TEXTURE_CUBE_MAP`.

For example, the following code binds the second descriptor in the `textures` array with the `GL_TEXTURE_2D` target:

```
glBindTexture(textures[1], GL_TEXTURE_2D);
```

Only one texture descriptor can be bound to a target at a time. In this example, any further operations involving the `GL_TEXTURE_2D` target will affect the texture represented by the `textures[1]` descriptor.

The `glPixelStorei` function identifies how pixel data should be loaded and stored. This associates a loading/storage parameter with an integer value. Table 10.12 lists the different parameters available.

Table 10.12

Loading/Storage Parameters of glPixelStorei

Parameter	Description
GL_UNPACK_ROW_LENGTH	The number of pixels per row when loading
GL_UNPACK_IMAGE_HEIGHT	The number of rows (if greater than 0)
GL_UNPACK_SKIP_ROWS	The number of rows to skip before loading
GL_UNPACK_SKIP_PIXELS	The number of pixels to skip before loading
GL_UNPACK_SKIP_IMAGES	The number of images to skip before loading
GL_UNPACK_ALIGNMENT	Byte alignment for loading each row (1, 2, 4, or 8)
GL_PACK_ROW_LENGTH	The number of pixels per row when storing
GL_PACK_SKIP_ROWS	The number of rows to skip before storing
GL_PACK_SKIP_PIXELS	The number of pixels to skip before storing
GL_PACK_ALIGNMENT	Byte alignment for storing each row (1, 2, 4, or 8)

The six parameters starting with `GL_UNPACK` identify how pixels should be unpacked from memory. For example, `GL_UNPACK_ROW_LENGTH` specifies the number of pixels in a row. If an image contains 128 pixels per row, this can be indicated with the following code:

```
glPixelStorei(GL_UNPACK_ROW_LENGTH, 128);
```

The four parameters starting with `GL_PACK` identify how pixels should be packed into memory. For example, `GL_PACK_ALIGNMENT` specifies the byte boundaries that should be employed when storing rows. The default value is 4.

`glTexParameteri` is like `glPixelStorei`, but instead of identifying how the image data should be loaded and stored, it configures how a texture should be displayed. Table 10.13 lists the parameter names accepted by `glTexParameteri`.

Table 10.13

Texture Display Parameters of glTexParameteri

Parameter	Characteristic
GL_TEXTURE_MIN_FILTER	The interpolation method used for minimization
GL_TEXTURE_MAG_FILTER	The interpolation method used for enlargement
GL_TEXTURE_WRAP_S	How the texture should be displayed in the s-direction at coordinates beyond its dimensions

GL_TEXTURE_WRAP_T	How the texture should be displayed in the t-direction at coordinates beyond its dimensions
GL_TEXTURE_WRAP_R	How the texture should be displayed in the r-direction at coordinates beyond its dimensions
GL_TEXTURE_MIN_LOD	The texture's lowest mipmap level
GL_TEXTURE_MAX_LOD	The texture's highest mipmap level
GL_TEXTURE_BASE_LEVEL	The index of the lowest-defined mipmap level
GL_TEXTURE_MAX_LEVEL	The index of the highest-defined mipmap level
GL_TEXTURE_SWIZZLE_R	Sets the value of the texel's R component
GL_TEXTURE_SWIZZLE_G	Sets the value of the texel's G component
GL_TEXTURE_SWIZZLE_B	Sets the value of the texel's B component
GL_TEXTURE_SWIZZLE_A	Sets the value of the texel's A component
GL_DEPTH_STENCIL_TEXTURE_MODE	How to read data from depth/stencil textures
GL_TEXTURE_COMPARE_MODE	Texture comparison mode for depth textures
GL_TEXTURE_COMPARE_FUNC	The comparison used in the depth texture test

The following discussion presents most of these parameters by dividing them into three categories: filtering modes, external coordinates, and level of detail. The topics of texture swizzling and texture-based depth testing are beyond the scope of this book.

Filtering Modes

Fragment shaders use coordinates to identify what point of a texture should be sampled. If a point lies between texture elements, or *texels*, the shader has two options. It can read data from the texel whose center is nearest or it can compute a weighted sum of the surrounding texels. These options are represented by two constants:

- GL_NEAREST — select data from the texel whose center is nearest
- GL_LINEAR — compute a weighted average of surrounding texels

If mipmaps aren't used, the GL_TEXTURE_MIN_FILTER and GL_TEXTURE_MAG_FILTER fields must be set to one of these two values. For example, if GL_TEXTURE_MIN_FILTER is set to GL_NEAREST, the renderer will use nearest-neighbor interpolation whenever it needs to minify texels. If GL_TEXTURE_MAX_FILTER is set to GL_LINEAR, the renderer will use bilinear interpolation when it needs to magnify texels in an image.

If mipmaps are used, there are more filtering modes to choose from. These include GL_NEAREST_MIPMAP_NEAREST, GL_LINEAR_MIPMAP_NEAREST, GL_NEAREST_MIPMAP_LINEAR, and GL_LINEAR_MIPMAP_LINEAR.

External Coordinates

If a shader tries to read data beyond a texture's boundary, a question arises regarding what value should be returned. An application can configure this by setting values for `GL_TEXTURE_WRAP_S`, `GL_TEXTURE_WRAP_T`, and `GL_TEXTURE_WRAP_R`. In texture space, the (s, t, r) axes correspond to the (x, y, z) axes in regular space. `glTexParameteri` can assign each of these parameters to one of four values:

- `GL_REPEAT` — In-range coordinates are extended beyond the texture. If the maximum dimension is N, an out-of-range texel X is set to the texel at X mod N
- `GL_MIRRORED_REPEAT` — Texels beyond the texture are set to the reflections of the corresponding in-range texels
- `GL_CLAMP_TO_EDGE` — Texels beyond the texture are set to the texels at the texture's edge
- `GL_CLAMP_TO_BORDER` — Texels beyond the texture are set to a border color

On the following page, the images in Figure 10.6 show what happens when a shader reads texels outside of the texture's boundaries. In each case, the coordinates run from −0.25 to 1.25.

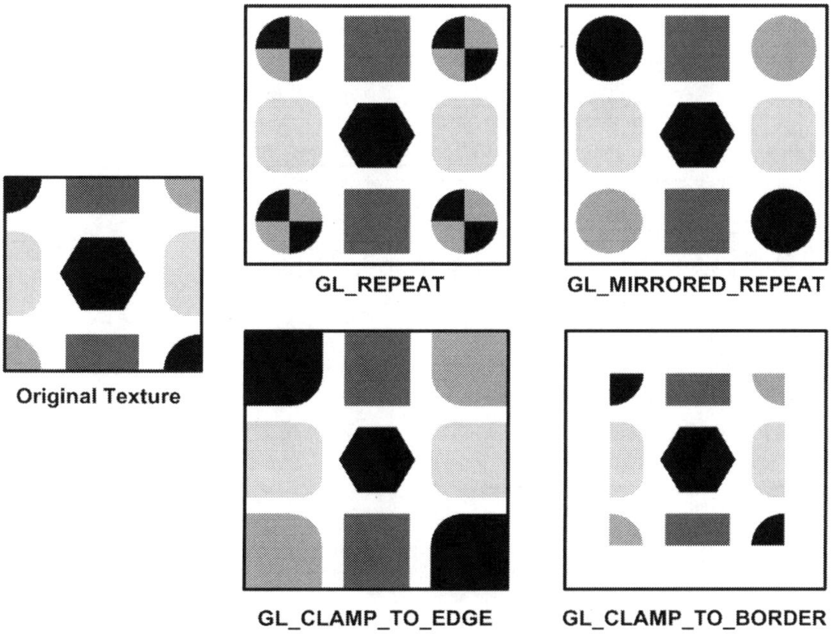

Figure 10.6 Using Texture in the ch10_texture_squares Project

In the figure, the GL_SAMPLER_ADDRESS_MODE_CLAMP_TO_BORDER example sets the border color to white. In code, this can be configured by calling glTexParameterfv. This is similar to glTexParameteri, but instead of setting a parameter equal to an integer, it sets the parameter to an array of floating-point values. To set the border color, the parameter name should be set to GL_TEXTURE_BORDER_COLOR and the array should contain the color's components.

The following code shows how this works. In this case, colorArray contains four floats that identify the color's R, G, B, and A components.

```
GLfloat colorArray[] = {1.0f, 1.0f, 1.0f, 1.0f};
glTexParameterfv(GL_TEXTURE_2D, GL_TEXTURE_BORDER_COLOR,
  colorArray);
```

Level of Detail and Mipmapping

As a viewer moves away from a texture, the texture's resolution should decrease. One way to accomplish this is by reading fewer samples from the texture. This is referred to as a level of detail (LOD) method. An application can specify the minimum LOD by calling glTexParameteri with the GL_TEXTURE_MIN_LOD parameter and the maximum LOD with the GL_TEXTURE_MAX_LOD parameter.

A common LOD method is to add mipmap levels to a texture, where each level's image has a lower resolution than the one preceding it. That is, the Level 0 image has the highest resolution, the Level 1 image has the next highest, and so on. An application can set the minimum mipmap level by calling glTexParameteri with the GL_TEXTURE_BASE_LEVEL parameter and the maximum mipmap level with the GL_TEXTURE_MAX_LEVEL parameter.

10.9.2 Associating Textures with Image Data

After the texture is created and its characteristics are set, an application can associate the texture with pixel data by calling glTexImage2D. This serves a similar purpose to glBufferData, but in addition to the data dimensions, glTexImage2D also needs to know how the pixels in the texture are structured. Three of the function's arguments provide this information:

- intformat identifies the number of components in each texel. Example values include GL_RGBA8, GL_RGBA16, GL_RGB8, GL_RGB16, GL_ALPHA8, GL_ALPHA16, GL_LUMINANCE8, and GL_LUMINANCE16.
- format identifies the properties of individual texels. Example values include GL_RGBA, GL_RGBA, GL_ALPHA, and GL_LUMINANCE.

- `type` sets the data type of each component. Example values include `GL_FLOAT`, `GL_INT`, `GL_UNSIGNED_BYTE`, and `GL_SHORT`.

To demonstrate how `glTexImage2D` can be used, the following code associates a two-dimensional texture with a 60-by-40 array of texels called `texData`. Each texel in `tex_data` is composed of RGB components and each component is an unsigned byte.

```
glTexImage2D(GL_TEXTURE_2D, 0, GL_RGB8, 60, 40, 0, GL_RGB,
  GL_UNSIGNED_BYTE, texData);
```

`glTexImage2D` is the last of the functions that need to be called to configure a texture. Once this is accomplished, the next step involves matching the coordinates of the texture to vertices in the model. This matching process is called *texture mapping*, and will be discussed next.

10.9.3 Texture Mapping in the Vertex Shader

In the ch10_three_squares project, the vertex shader received the color of each vertex and passed it to the fragment shader as an output variable. In addition to passing colors, a vertex shader can pass texture coordinates to the fragment shader that identify which location in the texture should be associated with the currently-processed vertex.

If you look at the code in the ch10_texture_squares project, you'll see that the texture coordinates are set with the following code:

```
// Texture coordinates
GLfloat tex_coords[] = {0.0f, 0.0f,
                        0.0f, 1.0f,
                        1.0f, 1.0f,
                        1.0f, 0.0f};
```

These values are stored in a VBO and passed to the vertex shader, which accesses them through its `in_texcoords` attribute. The following code declares `in_texcoords` and `new_texcoords` as vectors containing two floating point values:

```
layout (location = 1) in vec2 in_texcoords;
out vec2 new_texcoords;
```

In order to pass the texture coordinates to the fragment shader, the vertex shader needs to set the output variable `new_texcoords` equal to `in_texcoords`. This is accomplished with the following code:

```
new_texcoords = in_texcoords;
```

When the fragment shader receives the texture coordinates, it will use them to access the texture and set the fragment's color to the pixel data at the given location. This is discussed next.

10.9.4 Applying Textures in the Fragment Shader

In the ch10_three_squares project, the vertex shader passed color data using a variable called `new_color`. In the ch10_texture_squares project, the vertex shader passes texture coordinates using a variable called `new_texcoords`. The fragment shader receives this as an input variable called `new_texcoords`.

Fragment shaders can call a function named `texture`, which accesses the color at a given location in the currently-bound texture. This can be called in a number of different situations, but for a simple two-dimensional texture, its signature is given as follows:

```
vec4 texture(sampler2D sampler, vec2 coords)
```

The first argument is a *sampler*, which identifies the interpolation method to be used for the texture and how to display the texture when coordinates go beyond the texture's borders.

For basic sampling, there's no need to create a custom sampler. A fragment shader can access a built-in sampler as a *uniform*. Uniforms are similar to attributes in that shaders can access them as input. Unlike attributes, uniforms are constant. That is, they don't change from pixel to pixel or from fragment to fragment. The following code shows how the fragment shader accesses a two-dimensional sampler named sam:

```
uniform sampler2D sam;
```

The fragment shader uses `sam` in the `texture` function, which accesses the color at the location given by `new_texcoords`. This is accomplished in the following code:

```
vec3 color = texture(sam, new_texcoords).rgb;
```

Once the color is accessed, the fragment shader sets the color of the current fragment with the following code:

```
new_color = vec4(color, 1.0);
```

10.9.5 Code Discussion

The code in the ch10_texture_squares project closely resembles the code in the ch10_three_squares project discussed earlier. Both create three VAOs (one for each square) and each VAO is associated with two VBOs. But the ch10_texture_squares code is different in three important ways:

1. The second VBO contains texture coordinates for the vertices instead of color components.
2. The `main` function calls `initTextures` to allocate memory for the textures and associate each texture with data.
3. The `display` function calls `glBindTexture` to make a texture active. When a texture is active, the fragment shader will read pixels from that texture.

For the second point, the `initTextures` function starts by enabling two-dimensional texture processing and then allocates memory for three textures. This initializes the three descriptors in the `textures` array.

```
glEnable(GL_TEXTURE_2D);
glGenTextures(NUM_SQUARES, textures);
```

Once the descriptors are available, `initBuffers` iterates through them, binding each texture to the `GL_TEXTURE_2D` target and configuring its properties. Then it reads the pixels from the corresponding PNG using the `stbi_load` function discussed in Chapter 9. When the pixel data is available, `initBuffers` associates it with the currently-bound texture using the following code:

```
glTexImage2D(GL_TEXTURE_2D, 0, GL_RGB8, width[i], height[i],
  0, GL_RGB, GL_UNSIGNED_BYTE, texPixels);
```

For the third point, each of the three squares needs to be associated with a different texture. For this reason, the `display` function calls `glBindTexture` with a different texture descriptor before calling `glDrawArrays`. For example, to draw the second square, the following code makes the second VAO active and then binds the second texture descriptor.

```
glBindVertexArray(vao[1]);
glBindTexture(GL_TEXTURE_2D, textures[1]);
glDrawArrays(GL_TRIANGLE_FAN, 0, 4);
```

10.10 Summary

Before you can understand how CUDA accelerates graphical rendering, you need to understand how the rendering process works. Thankfully, the principles underlying OpenGL development aren't significantly different from those underlying CUDA development. In both cases, the host application initializes the data and the device processes it. The main difference is that OpenGL data is graphics-oriented and the device uses the processed data to draw pixels on a screen.

In an OpenGL application, the host code has to perform four main tasks. First, it needs to create vertex buffer objects (VBOs) from vertex data. Then it has to define attributes that identify the format and content of the VBO data. Shaders process this data, but before they can execute, the host needs to compile the shader code and deploy the linked program to the device. The last responsibility of the host is to launch the rendering process.

Once the rendering starts, the shaders handle the data processing. The vertex shader receives the attributes from the host and determines the final location of each incoming vertex. The fragment shader determines the final color of each pixel drawn in the display. Both shaders are coded in the OpenGL Shading Language, or GLSL. GLSL bears a number of similarities to CUDA, and both languages provide vector data types and mathematical functions to operate on the vectors.

The GL Utility Toolkit, or GLUT, creates windows capable of displaying the rendering produced by OpenGL. GLUT doesn't provide as many features as professional toolkits like Qt or GTK, but it's easy to learn and use. After you call the functions to initialize the window, you can tell GLUT which functions should respond to events. To configure GLUT to work with OpenGL, the `reshape` callback function should call OpenGL's `glViewport` function. The `display` callback function should call `glDrawArrays`.

Textures make it possible to add detail to the rendering. A texture's pixels, called texels, are used to cover a surface of the model. Coding with textures consists of three steps. First the host application creates texture objects and associates each with texels. Then the vertex shader matches the texture coordinates with vertices. Lastly, the fragment shader accesses the texture to determine which color to set as output.

Chapter 11

CUDA-OpenGL Interoperability

One of CUDA's most compelling capabilities is its ability to accelerate OpenGL-based rendering. This is especially valuable for applications that carry out computation on the CPU prior to visualization. For instance, a physics application might solve kinematic equations on the CPU to determine object motion, then leverage the GPU to efficiently perform the transformations and rendering required to visually represent that motion.

By integrating CUDA into an OpenGL application, we can place a greater portion of the computational load on the GPU, freeing the CPU to perform other tasks. For example, we can code kernels that solve kinematic equations without the CPU's involvement. The advantages are significant: not only is the GPU better suited for many types of computation but GPU-centric computing makes it unnecessary to transfer data between the CPU and GPU.

Another reason to use CUDA in rendering applications is that CUDA kernels are more flexible than OpenGL shaders:

- Kernels can invoke a wider number of functions.
- Kernels can share data among threads in a thread block using shared memory.
- Kernels can call synchronization functions that control when threads in a thread block execute.

The second and third points are important. A vertex shader can access only one vertex at a time and a fragment shader can access only one fragment at a time. But a CUDA kernel can access all the global and constant memory on a device, including the shared memory associated with its thread block. It can also synchronize threads in a thread block, enabling cooperative execution.

The disadvantage of combining CUDA and OpenGL is that programming becomes more complex. In addition to dealing with the CUDA data structures (buffers, streams, and so on), you have to deal with OpenGL structures like vertex buffer objects and vertex array objects.

To simplify development, this chapter presents a method of structuring code into stages. The two example applications in this chapter demonstrate how this method can be put into practice. But before we get to the example code, it's important to discuss the functions that establish interoperability between CUDA and OpenGL.

11.1 Overview

The fundamental concept underlying CUDA-OpenGL interoperability is that OpenGL data structures, such as VBOs and textures, can be accessed as regular CUDA buffers. This makes it possible for CUDA kernels to process OpenGL structures as though they were blocks of device memory.

It's important to see that data isn't transferred between OpenGL structures and CUDA buffers. The same region of memory is accessed in two different ways. After the CUDA kernel processes the data, OpenGL rendering continues normally.

To be precise, the process of using CUDA-OpenGL interoperability consists of seven general steps:

1. Create OpenGL objects, such as vertex buffer objects (VBOs), textures, or renderbuffer objects.
2. Register each object with CUDA and obtain a handle to the registered object.
3. Use the handle to map each OpenGL object, making it accessible to CUDA.
4. For each OpenGL object, obtain a CUDA-accessible pointer to the device memory containing the object's data.
5. Pass the CUDA-accessible pointers to a kernel and execute the kernel, processing the OpenGL data.
6. Unmap each OpenGL object, enabling it to be accessed by OpenGL.
7. Render the OpenGL scene using the processed data in the OpenGL objects.

This section presents the functions needed to perform these steps. Keep in mind that a mapped OpenGL object can't be used for rendering until it is unmapped.

11.1.1 Registering OpenGL Objects

Before an OpenGL structure can be accessed in CUDA, it needs to be registered. The registration function depends on the type of OpenGL structure to be registered. If the OpenGL structure is a VBO, the function to call is `cudaGraphicsGLRegisterBuffer`. If the structure is a texture or a renderbuffer, the function to call is `cudaGraphicsGLRegisterImage`.

Both of these functions are declared in the cuda_gl_interop.h header file, which is provided in the CUDA Toolkit. This discussion looks at both functions.

Registering a Vertex Buffer Object (VBO)

If an application wants to process VBO data using CUDA before the vertex shader executes, it needs to register the VBO by calling `cudaGraphicsGLRegisterBuffer`. The function's signature is given as follows:

```
cudaGraphicsGLRegisterBuffer(
  struct cudaGraphicsResource **resource, GLuint vbo_descriptor,
  unsigned int flags)
```

The second argument is the descriptor of the VBO to be registered. The first argument is a handle that references a `cudaGraphicsResource` that represents the structure. A separate handle must be obtained for each VBO to be registered.

The last argument identifies how the OpenGL object will be accessed by CUDA. It can be set to one of three values:

1. `cudaGraphicsRegisterFlagsNone` — the VBO will be read from and written to by CUDA
2. `cudaGraphicsRegisterFlagsReadOnly` — CUDA will read from the VBO, but will not write to the VBO
3. `cudaGraphicsRegisterFlagsWriteDiscard` — CUDA will not read from the VBO, and write operations will overwrite the entire VBO

For example, if `vbo_desc` is a VBO descriptor, the following code registers the VBO with CUDA and provides CUDA with read/write access.

```
struct cudaGraphicsResource *cudaRes;
cudaGraphicsGLRegisterBuffer(&cudaRes, vbo_desc,
  cudaGraphicsRegisterFlagsNone);
```

Registering a Texture or Renderbuffer

If an application wants to process data in a texture or renderbuffer object using a CUDA kernel, it needs to register the object with CUDA by calling `cudaGraphicsGLRegisterImage`. The function's signature is given as follows:

```
cudaGraphicsGLRegisterImage(
  struct cudaGraphicsResource **resource,
  GLuint image, GLenum target, unsigned int flags)
```

The arguments are similar to those of the `cudaGraphicsGLRegisterImage` function discussed earlier. The second argument is the descriptor of the OpenGL structure and the third argument identifies the target bound to the structure. The target can be set to `GL_TEXTURE_2D`, `GL_TEXTURE_RECTANGLE`, `GL_TEXTURE_CUBE_MAP`, `GL_TEXTURE_3D`, `GL_TEXTURE_2D_ARRAY`, or `GL_RENDERBUFFER`.

The `flags` argument identifies how the OpenGL object will be accessed by CUDA. It can be set to one of five values:

1. `cudaGraphicsRegisterFlagsNone` — CUDA will read and write to the OpenGL structure
2. `cudaGraphicsRegisterFlagsReadOnly` — CUDA will read from the OpenGL structure, but will not write to the VBO
3. `cudaGraphicsRegisterFlagsWriteDiscard` — CUDA will not read from the VBO, and write operations will overwrite the entire OpenGL structure
4. `cudaGraphicsRegisterFlagsSurfaceLoadStore` — CUDA will bind this resource to a surface reference
5. `cudaGraphicsRegisterFlagsTextureGather` — CUDA will perform texture gather operations on this resource

For example, if `tex` represents a texture, the following code registers the texture with CUDA and specifies that the CUDA kernel should only read from the texture:

```
struct cudaGraphicsResource *cudaRes;
cudaGraphicsGLRegisterImage(&cudaRes, tex, GL_TEXTURE_2D,
  cudaGraphicsRegisterFlagsReadOnly);
```

This chapter presents example projects that access vertex buffer objects in CUDA. Chapter 12 will present a project that registers a texture with CUDA.

11.1.2 Mapping and Unmapping Resources

After an OpenGL object has been registered, the application needs to grant CUDA access to the object. This is called *mapping* the object, and it's accomplished by calling `cudaGraphicsMapResources`, whose signature is given as follows:

```
cudaGraphicsMapResources(int count,
  cudaGraphicsResource **resources, cudaStream_t stream = 0)
```

The first argument identifies the number of OpenGL objects to map and the second identifies the `cudaGraphicsResource` obtained through registration. The last argument identifies the CUDA stream that should be used for the mapping operation, and Chapter 5 discussed streams in detail. If this argument is set to 0, the operation will run in the default stream.

In order for a CUDA kernel to access a mapped OpenGL object, the application needs to obtain a pointer to device memory similar to that provided by `cudaMalloc`. The function that provides this pointer is `cudaGraphicsResourceGetMappedPointer`, and its signature is given as follows:

```
cudaGraphicsResourceGetMappedPointer(void** devPtr, size_t* size,
  cudaGraphicsResource *resource)
```

The first argument points to the block of device memory containing the OpenGL data, and can be passed to a CUDA kernel like a regular argument. The second argument contains the size of the memory block, and the last argument references the `cudaGraphicsResource` structure provided by registration.

A mapped OpenGL object can't be used in rendering, so applications need to unmap each object after CUDA processing has finished. Object unmapping is performed by the `cudaGraphicsUnmapResources` function, which is given as follows:

```
cudaGraphicsUnmapResources(int count,
  cudaGraphicsResource_t* resources, cudaStream_t stream = 0)
```

The arguments here are the same as those used in `cudaGraphicsMapResources`. The first argument identifies the number of OpenGL objects to be unmapped and the second argument references the `cudaGraphicsResource` obtained during the registration process.

An example will clarify how the registration, mapping, and unmapping processes work. The following code shows how these steps can be performed.

```
// Step 1: Register the VBO
cudaGraphicsResource* cudaRes;
cudaGraphicsGLRegisterBuffer(&cudaRes, vbo,
  cudaGraphicsMapFlagsNone);

// Step 2: Map the VBO so CUDA can access it
cudaGraphicsMapResources(1, &cudaRes, 0);

// Step 3: Get device pointer to the mapped resource
float* vboMem;
size_t num_bytes;
cudaGraphicsResourceGetMappedPointer((void**)&vboMem,
  &num_bytes, cudaRes);

// Step 4: Launch kernel that reads/writes devPtr
foo<<<grid, block>>>(vboMem);

// Step 5: Unmap so OpenGL can use it again
cudaGraphicsUnmapResources(1, &cudaRes, 0);
```

In Step 3 the `cudaGraphicsResourceGetMappedPointer` initializes `vboMem` so that it can serve as a regular CUDA buffer. Then in Step 4, `vboMem` is passed to the kernel (`foo`) like a regular kernel parameter. When the VBO is unmapped in Step 5, `vboMem` can no longer be accessed by CUDA.

11.2 Simple Example

The ch11_basic_interop project provides a simple example of how CUDA-OpenGL interoperability can be implemented. The project renders the same model as the ch10_three_squares project and uses the same shaders, but it executes a CUDA kernel to initialize the vertex buffer objects before rendering.

In code, the primary difference is that the `main` function in the ch11_basic_interop source code calls a function called `runCuda`. This updates the content of the application's VBOs by performing five operations:

1. Calls `cudaGraphicsGLRegisterBuffer` to register each of the six VBOs.
2. Maps OpenGL resources for CUDA access by calling `cudaGraphicsMapResources`.
3. Invokes `cudaGraphicsResourceGetMappedPointer` to obtain a pointer to device memory for each of the vertex buffer objects.

4. Calls the `initVBOs` kernel with one thread, passing it the pointers to the vertex buffer objects.
5. Frees the OpenGL objects for rendering by calling `cudaGraphicsUnmapResources`.

Listing 11.1 presents the code of the `runCuda` function in the ch11_basic_interop/main.cu source file.

Listing 11.1 ch11_basic_interop/main.cu - runCuda Function

```
// Perform CUDA operations
void runCuda() {

  int i;

  // Register the OpenGL objects
  for (i = 0; i < NUM_VBOS; i++) {
    cudaGraphicsGLRegisterBuffer(&cudaRes[i], vbo[i],
      cudaGraphicsRegisterFlagsWriteDiscard);
  }

  // Map the OpenGL resources
  cudaGraphicsMapResources(NUM_VBOS, cudaRes, 0);

  // Create memory objects from the VBOs
  float4* floatMem[NUM_VBOS];
  size_t memSize[NUM_VBOS];
  for (i = 0; i < NUM_VBOS; i++) {

    // Initialize the device memory pointer
    cudaGraphicsResourceGetMappedPointer(
      reinterpret_cast<void**>(&floatMem[i]),
      &memSize[i], cudaRes[i]);
  }

  // Invoke the kernel
  initVBOs<<<1, 1>>> (floatMem[0], floatMem[1],
    floatMem[2], floatMem[3], floatMem[4], floatMem[5]);

  // Unmap the OpenGL resources
  cudaGraphicsUnmapResources(NUM_VBOS, cudaRes, 0);
}
```

This code invokes the kernel named `initVBOs`, whose code is given in Listing 11.2. This sets the values in all six of the application's vertex buffer objects.

Listing 11.2 ch11_basic_interop/main.cu - initVBOs Kernel

```
__global__ void initVBOs(float4* vbo0, float4* vbo1,
  float4* vbo2, float4* vbo3, float4* vbo4, float4* vbo5) {

  // Initialize the first VBO
  vbo0[0] = make_float4(-0.15f, -0.15f, 1.00f, -0.15f);
  vbo0[1] = make_float4(0.15f, 1.00f, 0.15f, 0.15f);
  vbo0[2] = make_float4(1.00f, 0.15f, -0.15f, 1.00f);

  // Initialize the second VBO
  vbo1[0] = make_float4(0.00f, 0.00f, 0.00f, 0.25f);
  vbo1[1] = make_float4(0.00f, 0.00f, 0.50f, 0.00f);
  vbo1[2] = make_float4(0.00f, 0.75f, 0.00f, 0.00f);

  // Initialize the third VBO
  vbo2[0] = make_float4(-0.30f, -0.30f, 0.00f, -0.30f);
  vbo2[1] = make_float4(0.30f, 0.00f, 0.30f, 0.30f);
  vbo2[2] = make_float4(0.00f, 0.30f, -0.30f, 0.00f);

  // Initialize the fourth VBO
  vbo3[0] = make_float4(0.00f, 0.00f, 0.00f, 0.00f);
  vbo3[1] = make_float4(0.25f, 0.00f, 0.00f, 0.50f);
  vbo3[2] = make_float4(0.00f, 0.00f, 0.75f, 0.00f);

  // Initialize the fifth VBO
  vbo4[0] = make_float4(-0.45f, -0.45f, -1.00f, -0.45f);
  vbo4[1] = make_float4(0.45f, -1.00f, 0.45f, 0.45f);
  vbo4[2] = make_float4(-1.00f, 0.45f, -0.45f, -1.00f);

  // Initialize the sixth VBO
  vbo5[0] = make_float4(0.00f, 0.00f, 0.00f, 0.00f);
  vbo5[1] = make_float4(0.00f, 0.25f, 0.00f, 0.00f);
  vbo5[2] = make_float4(0.50f, 0.00f, 0.00f, 0.75f);
}
```

Ideally, it wouldn't be necessary to pass a separate argument to the kernel for each VBO to be processed. But the `cudaGraphicsResourceGetMappedPointer` function returns one pointer for each object, so the buffers can't be accessed as a single contiguous block of memory. If possible, it's a good idea to combine all the VBO data into a single object.

When the application terminates, GLUT calls the `close` function to free resources. This unregisters CUDA resources by calling `cudaGraphicsUnregisterResource` in the following way:

```
// Unregister CUDA resources
for (int i = 0; i < NUM_VBOS; i++) {
    cudaGraphicsUnregisterResource(cudaRes[i]);
}
```

Calling this function decreases the reference count of the CUDA resource and prevents memory leaks. After all the resources are unregistered, the application deletes the vertex buffer objects by calling `glDeleteBuffers`. Then it calls `glutLeaveMainLoop`, which tells GLUT to stop processing.

11.3 Animation

If a model remains static, it doesn't matter if the rendering is performed by a GPU or a CPU. But when thousands of pixels need to be updated with each frame, it becomes vital to perform as much processing on the GPU as possible. CUDA can help make this possible, and the second example of CUDA-OpenGL interoperability will implement animation.

Specifically, this section explores the ch11_spinning_sphere project, which draws a sphere that spins around its vertical axis. On the following page, Figure 11.1 shows what it looks like.

If you compare the code in the ch11_spinning_sphere project with the code in the ch11_basic_interop project, you'll see that the overall application structure is the same, and the `main` function invokes many of the same functions, including `initBuffers`, `initShaders`, `runCuda`, and `display`.

But the code in the ch11_spinning_sphere project has four significant differences:

1. All the vertex data is contained in a single VBO that contains 1,024 `float`s.
2. The `display` function is called repeatedly to animate the model.
3. With each call, `display` executes the CUDA kernel to compute new vertex coordinates.
4. The CUDA kernel computes the latitude and longitude of each point, and updates the vertex buffer object with the location of each vertex.

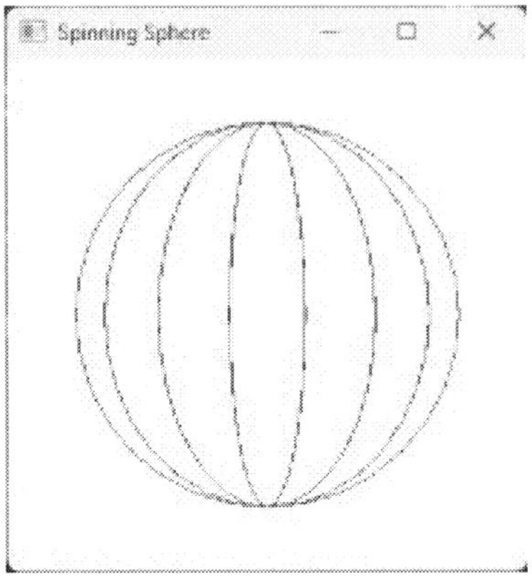

Figure 11.1 The Spinning Sphere

11.3.1 The Vertex Buffer Object

In the ch11_spinning_sphere project, all the vertex data is stored in a single vertex buffer object (VBO). This is because the model consists of a single object (the sphere) and only one color is used for rendering. This makes the `initBuffers` function very simple, and its code is given as follows:

```
void initBuffers() {

  // Create vertex array objects
  glGenVertexArrays(1, &vao);
  glBindVertexArray(vao);

  // Create vertex buffers
  glGenBuffers(1, &vbo);

  // VBO containing sphere coordinates
  glBindBuffer(GL_ARRAY_BUFFER, vbo);
  glBufferData(GL_ARRAY_BUFFER, 4 * NUM_VERTICES *
    sizeof(GLfloat), NULL, GL_DYNAMIC_DRAW);
  glVertexAttribPointer(0, 4, GL_FLOAT, GL_FALSE, 0, 0);
  glEnableVertexAttribArray(0);
}
```

It's important to see how `glBufferData` is called. The second argument sets the size of the VBO to `4 * NUM_VERTICES * sizeof(GLfloat)` bytes. `NUM_VERTICES` is set to 256, so the VBO occupies 4,096 bytes. The third argument is `NULL`, which means the object won't be initialized with data from the host.

11.3.2 Animation and Display

To keep the sphere spinning, the `display` function must be called repeatedly. This is made possible through the `glutPostRedisplay` function, which alerts the application that the window needs to be redrawn. In response, `display` is called again. The code for the `display` function is given as follows:

```
void display() {

  // Clear the rendering surface
  glClear(GL_COLOR_BUFFER_BIT | GL_DEPTH_BUFFER_BIT);

  // Launch kernel to update VBO
  runCuda();

  // Render model with updated VBO
  glBindVertexArray(vao);

  // Draw vertices in a line loop
  glDrawArrays(GL_LINE_LOOP, 0, NUM_VERTICES);

  // Update timing
  tick += 0.001f;

  // Unbind the vertex array object
  glBindVertexArray(0);

  // Make the framebuffer image visible
  glutSwapBuffers();

  // Restart display
  glutPostRedisplay();
}
```

After clearing the rendering state, the `display` function calls `runCuda`, which executes the kernel. After the kernel sets the vertices of the sphere, the `glDrawArrays` function renders the vertices in the window. `glDrawArrays` connects the vertices in a line loop, which means that it draws a line from each vertex to its successor, and the last vertex is connected to the first.

The `tick` variable is particularly important. It starts with a value of zero and each iteration of display increments it by 0.001. This new value is sent to the kernel, which uses it to update the coordinates of the sphere's vertices. This change in coordinates animates the sphere, and the increment controls how quickly the sphere rotates.

11.3.3 Executing the Kernel

To see how the kernel computes the sphere's vertices, you need to understand spherical coordinates. Normally, we identify points in space using rectangular coordinates (x, y, z). But for points on a sphere, it's more convenient to use (r, θ, φ):

- r — the radius of the sphere, the distance from the sphere's center to the point
- θ (theta) — the angle from the top of the sphere (the *zenith*) to the point
- φ (phi) — the angle along the circle perpendicular to the line connecting the sphere's center and zenith

Figure 11.2 shows how these spherical coordinates are related. The point P is located on the surface of a sphere whose center is C and whose zenith is Z. The coordinates of P are given as (r, θ, φ).

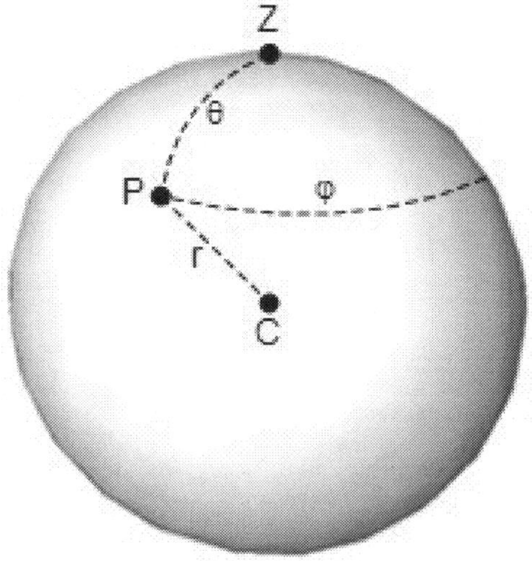

Figure 11.2 Spherical Coordinates

The angle θ runs from the zenith to the nadir (the lowest point on the sphere), so its angular measure runs from 0 to π radians. The angle φ runs the length of an entire circle, so its angular measure runs from 0 to 2π radians. The VBO contains 256 vertices divided into 16 longitudinal lines drawn between the zenith and the nadir. Each line contains 16 points and each point on a line has a different value of θ.

OpenGL accepts only rectangular coordinates for its vertices, so each (r, θ, φ) triple needs to be converted into an (x, y, z) triple. Using trigonometry, this conversion can be computed with three equations:

- $x = r \sin\theta \cos\varphi$
- $y = r \sin\theta \sin\varphi$
- $z = r \cos\theta$

The CUDA kernel uses these equations to compute the coordinates of each point on the sphere. There are 256 vertices in total and the radius is set to 0.75. The kernel code is given as follows:

```
// Initialize VBO data
__global__ void sphere(float4* vertices, float tick) {

  // Compute latitude and longitude
  int id = threadIdx.x;
  int longitude = id / 16;
  int latitude = id % 16;

  // Set the sphere's radius
  float radius = 0.75;

  // Compute spherical coordinates
  float sign = -2.0f * (longitude % 2) + 1.0f;
  float phi = 2.0f * CUDART_PI_F * longitude / 16 + tick;
  float theta = CUDART_PI_F * latitude / 16;

  // Compute rectangular coordinates
  vertices[id].x = radius * sin(theta) * cos(phi);
  vertices[id].y = radius * sign * cos(theta);
  vertices[id].z = radius * sin(theta) * sin(phi);
  vertices[id].w = 1.0f;
}
```

The vertices are divided into lines of 16 vertices each. The first sixteen vertices form the first line, the second sixteen vertices form the second line, and so on. The points in a line all have the same value of φ, and the `tick` variable, incremented by the `display` function, increases this value. As φ increases, the lines rotate counter-clockwise around the vertical axis. If the code is changed so that `tick` is decremented instead of incremented, the lines will rotate clockwise around the axis.

To access the `CUDART_PI_F` constant in the kernel, the source code includes the math_constants.h header file. No extra header files are needed to access the `sin` or `cos` functions.

11.4 Summary

Despite its age, OpenGL is frequently used in games, scientific visualization, and computer-aided design (CAD). The fact that this rendering can be accelerated with CUDA is an extraordinary capability. The goal of this chapter is to show how this acceleration can be implemented in code.

The central concept is *data sharing*. OpenGL and CUDA don't transfer data between buffers. Instead, they read and write to the same memory. To make this sharing possible, applications need to register and map OpenGL objects. Then the application can obtain pointers to device memory that can be accessed by CUDA kernels.

This chapter has presented two projects that demonstrate CUDA-OpenGL interoperability. The first executes a CUDA kernel that initializes the vertex coordinates and colors in a static rendering. The code's operation consists of four stages: initializing OpenGL objects, registering and mapping the objects, executing the kernel, and displaying the drawing.

The second application uses the same code structure to display an animated sphere. Animation is accomplished by executing the kernel repeatedly, each time with an updated value of the vertex angles. With each iteration, the kernel computes different vertex coordinates and the model rotates around its axis.

Chapter 12

Processing Textures and Surfaces

In the early 1990s, the computer game *Wolfenstein 3D* single-handedly launched the genre known as the first-person shooter (FPS). Since its release, the basic elements of FPS gameplay haven't changed: keystrokes move the character, mouse motion changes the character's viewpoint, and clicks fire the character's weapon.

Computer graphics, on the other hand, have changed dramatically. Instead of pixelated bad guys who look and move like LEGO men, monsters in modern games are rendered with incredible detail.

For the most part, these improvements have been made possible by *textures*. As discussed in Chapter 10, a texture is an image that the renderer stretches or shrinks to cover a surface. Simple applications apply textures to a model and don't make any changes. But for high-quality special effects, applications use advanced image processing techniques to update the texture in real time.

Using CUDA, applications can update textures before they're rendered to the screen. This makes it possible to implement advanced features like changing the lighting depending on the time of day or casting reflections from shiny surfaces. The process of using CUDA to update textures is similar to that of updating vertex buffers. Once a texture object is registered, it needs to be mapped for CUDA to process it. Once the processing is finished, the texture must be unmapped so that OpenGL can render it.

This chapter explains the functions that make texture processing with CUDA possible. It also presents the code in a project that filters a texture using the Gaussian blur. Chapter 9 explained how images can be filtered with NPPI, but in this example, the filter is implemented in a kernel that executes before the texture is rendered.

12.1 Gaussian Blur

Chapter 9 presented the Nvidia Performance Primitives related to image processing (NPPI) and showed how its functions can be used to filter images. The example project discussed in this chapter focuses on a specific type of image filter called the *Gaussian blur*. This smooths an image by applying a Gaussian function, which reduces noise and produces a soft look.

As discussed in Chapter 9, image filtering is performed using a mathematical process called *convolution*. This combines values in the filter's kernel with pixels of an image, and produces pixels of a new image. Figure 12.1 gives an idea of what convolution looks like for a 3-by-3 kernel.

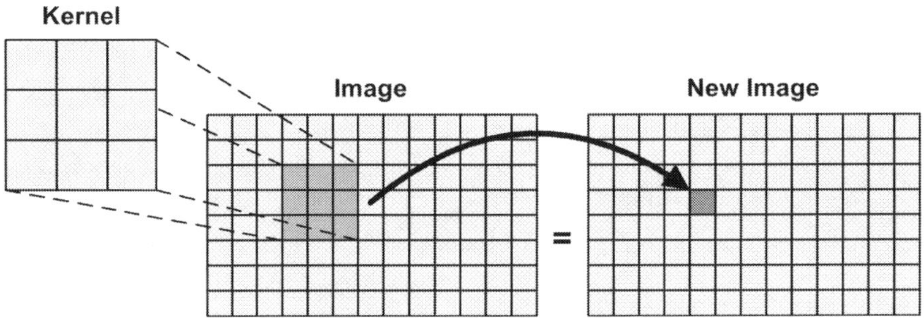

Figure 12.1 Image Filtering with Convolution

Convolution requires taking a two-dimensional dot product for each pixel in the image. To compute this dot product, the application needs to multiply the filter's values by pixels in the image and set the pixel of the new image equal to the sum of the products. On the next page, Figure 12.2 provides an example of how two-dimensional dot products are computed.

Using convolution, applications can produce filtered images with many impressive effects. Chapter 9 explained how averaging filters reduce noise in an image, and walked through the development of an application that increases an image's contrast using a sharpening filter.

In the world of artificial intelligence, image filters are frequently used for *edge detection*. These filters accentuate gradients in an image, and make it possible to locate boundaries of objects. After locating boundaries, applications can split an image into segments and examine each segment individually.

1.5	0.2	3.6
6.0	7.2	8.4
5.1	1.9	2.3

•

7.6	9.5	2.7
5.5	4.9	1.2
1.7	0.3	8.9

$$= 1.5(7.6) + 0.2(9.5) + 3.6(2.7) + 6.0(5.5) + 7.2(4.9) + 8.4(1.2) + 5.1(1.7) + 1.9(0.3) + 2.3(8.9) = 131.09$$

Figure 12.2 Two-Dimensional Dot Product

This discussion is concerned with an advanced filter called the Gaussian blur. Like the averaging filter, its goal is to remove noise from an image. But it produces a more natural looking result that preserves edges.

To understand the Gaussian blur, it helps to be familiar with the Gaussian curve, also known as the bell curve. This is used by statisticians to represent the graph of a normal distribution, with a rounded peak that tapers at each end. Figure 12.3 illustrates two Gaussian curves.

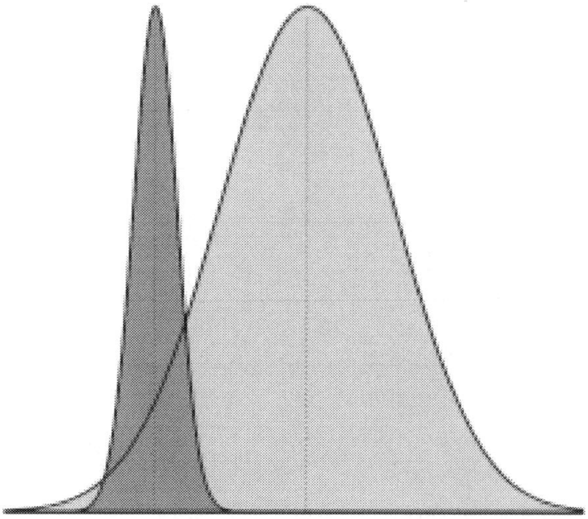

Figure 12.3 One-Dimensional Gaussian Curves

Mathematically, Gaussian curves are represented by the exponential function. If the mean is zero and the standard deviation is 1, the equation for a one-dimensional Gaussian curve is $f(x) = \exp(-x^2)$. The equation for a two-dimensional Gaussian curve is $f(x, y) = \exp(-x^2 - y^2)$. On the following page, Figure 12.4 shows what the two-dimensional Gaussian curve looks like.

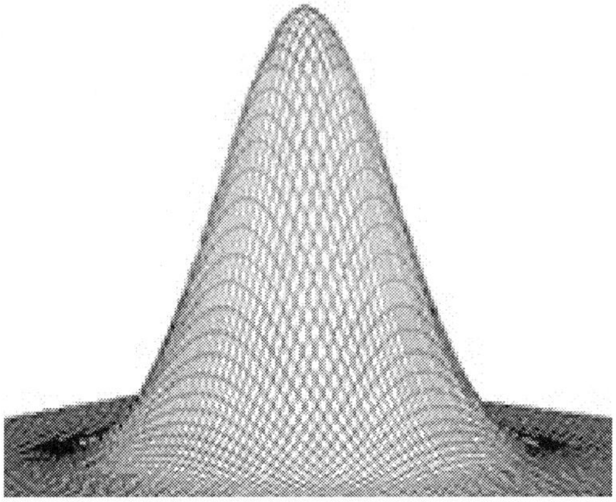

Figure 12.4 Two-Dimensional Gaussian Curve

To implement the two-dimensional Gaussian in a kernel, it's necessary to approximate its values in a matrix. The following 3-by-3 matrix presents one possible kernel for the Gaussian blur:

$$k_{Gaussian} = \frac{\begin{vmatrix} 1 & 2 & 1 \\ 2 & 4 & 2 \\ 1 & 2 & 1 \end{vmatrix}}{16}$$

Filtering with this kernel produces a blurring effect similar to that of the averaging filter, but it gives priority to the central pixel and the pixels adjacent to it. This means that the filtered image keeps more of the differences between one pixel and the next.

It's important to note that the kernel used for this filter is only one implementation of the Gaussian blur. The kernel's values can be modified to produce more or less pixel blurring, and the following matrix presents a 5-by-5 kernel for a better Gaussian blur.

$$k_{Gaussian} = \frac{\begin{vmatrix} 1 & 4 & 6 & 4 & 1 \\ 4 & 16 & 24 & 16 & 4 \\ 6 & 24 & 36 & 24 & 6 \\ 4 & 16 & 24 & 16 & 4 \\ 1 & 4 & 6 & 4 & 1 \end{vmatrix}}{256}$$

12.2 Textures and Surfaces

Chapter 4 discussed many of the data types that can be accessed by CUDA kernels, including scalar and vector types. But there are two important data types that weren't mentioned:

1. `cudaTextureObject_t` — a *texture*, which can be read from using configurable filtering and addressing modes
2. `cudaSurfaceObject_t` — a *surface*, which can be read from or written to

These data types weren't mentioned earlier because they're only used by applications that need to process image data. In general, I recommend using NPPI for GPU-based image manipulation. But if you need to perform custom operations on images that aren't supported by NPPI, it's a good idea to be aware of these data types and their usage.

This section looks at both of these data types in detail, and a later section will demonstrate how they can be used in code. But before you can provide data to textures and surfaces, you need to be familiar with CUDA arrays.

12.2.1 CUDA Arrays

In earlier chapters, applications called `cudaMalloc` to allocate blocks of device memory and `cudaMemcpy` to copy host data to the device. This is fine for one-dimensional data, but CUDA provides special functions for handling data with two or three dimensions.

This kind of data can be stored in *CUDA arrays*. Data in a two-dimensional image can be allocated with `cudaMallocArray`, while three-dimensional data can be allocated with `cudaMalloc3D` or `cudaMalloc3DArray`.

In this chapter, the goal is to process a two-dimensional image in a CUDA kernel. The signature of `cudaMallocArray` is given as follows:

```
cudaMallocArray(struct cudaArray** array,
  const struct cudaChannelFormatDesc *desc, size_t width,
  size_t height, unsigned int flags = 0)
```

The first argument identifies the block of allocated device memory for the image. The variable must be declared as a `cudaArray_t`.

The second argument describes the components of the image's pixels (called *channels*). In C++, the `cudaChannelFormatDesc` can be created by calling `cudaCreateChannelDesc` with a template type that sets the type of each channel.

For example, the following code states that each pixel contains a single channel whose value is given as a `float`:

```
cudaChannelFormatDesc desc = cudaCreateChannelDesc<float>();
```

If each pixel contains four channels (red, green, blue, and alpha), and each value is given as an unsigned character (0-255), the following descriptor should be used:

```
cudaChannelFormatDesc desc = cudaCreateChannelDesc<uchar4>();
```

The rest of the arguments in `cudaMallocArray` are straightforward. `width` sets the image's width, `height` sets the image's height, and `flags` can be set to one of two values:

- `cudaArrayDefault` — default array allocation
- `cudaArraySurfaceLoadStore` — allocates an array that can be read from or written to using a surface reference

To demonstrate this, the following code allocates memory for an image identified with a variable named `inputImage`. Each pixel of the image contains four channels whose values are given as unsigned characters. The image is 200 pixels wide and 100 pixels high.

```
cudaArray_t inputImage;
cudaChannelFormatDesc desc = cudaCreateChannelDesc<uchar4>();
cudaMallocArray(&inputImage, &desc, 200, 100);
```

If `cudaMallocArray` executes successfully, the application can copy data to the CUDA array. The `cudaMemcpy` function can't be used with CUDA arrays, so applications need to call one of the functions listed in Table 12.1.

Table 12.1

Copying Data to a CUDA Array

Name	Description
`cudaMemcpyToArray( struct cudaArray* dst, size_t wOffset, size_t hOffset, const void* src, size_t count, enum cudaMemcpyKind kind)`	Copies data to a CUDA array

`cudaMemcpyToArrayAsync( struct cudaArray* dst, size_t wOffset, size_t hOffset, const void* src, size_t count, enum cudaMemcpyKind kind, cudaStream_t stream = 0)`	Copies data to a CUDA array in the given stream
`cudaMemcpyArrayToArray( struct cudaArray* dst, size_t wOffsetDst, size_t hOffsetDst, const struct cudaArray* src, size_t wOffsetSrc, size_t hOffsetSrc, size_t count, enum cudaMemcpyKind kind = cudaMemcpyDeviceToDevice)`	Copies data from one CUDA array to another
`cudaMemcpy2DToArray( struct cudaArray* dst, size_t wOffset, size_t hOffset, const void* src, size_t spitch, size_t width, size_t height, enum cudaMemcpyKind kind)`	Copies two-dimensional data to a CUDA array
`cudaMemcpy2DToArrayAsync( struct cudaArray* dst, size_t wOffset, size_t hOffset, const void* src, size_t spitch, size_t width, size_t height, enum cudaMemcpyKind kind, cudaStream_t stream = 0)`	Copies two-dimensional data to a CUDA array in the given stream
`cudaMemcpy2DArrayToArray( struct cudaArray* dst, size_t wOffsetDst, size_t hOffsetDst, const struct cudaArray* src, size_t wOffsetSrc, size_t hOffsetSrc, size_t width, size_t height, enum cudaMemcpyKind kind = cudaMemcpyDeviceToDevice)`	Copies data from a two-dimensional array to a CUDA array

To transfer image data to a CUDA array, the code in the ch12_gaussian_blur project uses `cudaMemcpy2DToArray`. The `wOffset` and `hOffset` parameters identify the byte offset of the pixel in the upper left corner, and `spitch` identifies the number of bytes separating the start of one row and the start of the following row.

The last parameter identifies the direction of the copy. Like the last parameter of `cudaMemcpy`, this can be set to `cudaMemcpyHostToHost`, `cudaMemcpyHostToDevice`, `cudaMemcpyDeviceToHost`, or `cudaMemcpyDeviceToDevice`.

For example, suppose the image data is stored in a region identified with `inputData`, and has no padding. If the image's size is 200-by-100 and the upper-left pixel is located at (0, 0), the following code will copy the data from `inputData` to `inputArray`:

```
// Set image dimensions
size_t width = 200, height = 100;

// Set image pitch
size_t pitch = width * sizeof(uchar4);

// Copy image data from host to device
cudaMemcpy2DToArray(inputArray, 0, 0, inputData, pitch,
  pitch, height, cudaMemcpyHostToDevice)
```

It's important to notice that the `width` parameter is provided in bytes but the `height` parameter is set equal to the number of rows. Because there is no padding, the width can be set equal to the pitch.

12.2.2 Creating Textures

Chapter 10 explained how OpenGL makes it possible to create textures and access them in fragment shaders. CUDA provides its own data structures for textures, and while there are similarities between OpenGL textures and CUDA textures, there are also significant differences.

On an Nvidia GPU, data in a texture is accessed differently than other global data. There are three crucial differences:

1. Texture data is *read-only*.
2. Textures are accessed through a dedicated cache that stores data based on spatial location.
3. Texture data is read using a *sampler*, which can be configured to use filtering methods, addressing methods, and coordinate modes.

In code, CUDA textures are represented by cudaTextureObject_t objects and they can be created with the cudaCreateTextureObject function. Its signature is given as follows:

```
cudaCreateTextureObject(cudaTextureObject_t* texObj,
  const cudaResourceDesc* resDesc, const cudaTextureDesc* txDesc,
  const cudaResourceViewDesc* resViewDesc )
```

The first argument is a pointer to the initialized cudaTextureObject_t object. The next three arguments are descriptors that provide information about the texture's data and the manner in which the data should be accessed. The following discussion looks at each of these descriptor structures.

The Resource Descriptor

The cudaResourceDesc structure identifies the nature of the data from which the texture should be created. This has two fields, and the second field depends on the source of the texture's data. To grasp this, it helps to look at the structure's definition:

```
struct cudaResourceDesc {
  enum cudaResourceType resType;
  union {
    struct {
      cudaArray_t array;
    } array;
    struct {
      cudaMipmappedArray_t mipmap;
    } mipmap;
    struct {
      void *devPtr;
      struct cudaChannelFormatDesc desc;
      size_t sizeInBytes;
    } linear;
    struct {
      void *devPtr;
      struct cudaChannelFormatDesc desc;
      size_t width;
      size_t height;
      size_t pitchInBytes;
    } pitch2D;
  } res;
```

The first field, resType, identifies the source of the data from which the texture is created. This can be set to one of four values.

- cudaResourceTypeArray — data comes from a CUDA array (cudaArray_t)
- cudaResourceTypeMipmappedArray — data comes from a mipmap array (cudaMipmappedArray_t)
- cudaResourceTypeLinear — data comes from linear device memory (void*)
- cudaResourceTypePitch2D — data comes from 2-D device memory (void*)

The example code in this chapter creates a texture from a CUDA array. So we'll initialize the cudaResourceDesc using code like the following:

```
struct cudaResourceDesc resDesc;
resDesc.resType = cudaResourceTypeArray;
resDesc.array = inputImage;
```

In this code, the array field is set to a variable named inputImage. This should reference memory allocated with a function like cudaMallocArray.

The Texture Descriptor

The third argument of cudaCreateTextureObject is a cudaTextureDesc that identifies how the texture should be accessed. Table 12.2 lists the fields of this structure.

Table 12.2

Fields of the cudaTextureDesc Structure

Field	Type	Description
addressMode	cudaTextureAddressMode[3]	Specifies how texels should be computed for out-of-range coordinates
filterMode	cudaTextureFilterMode	Identifies how texels should be computed when minifying or magnifying an image
readMode	cudaTextureReadMode	Sets the data type for texels read from the texture
sRGB	int	Specifies if sRGB data should be linearized
borderColor	float[4]	Sets the color of texels on the border depending on the addressMode field

normalizedCoords	int	Identifies if texture coordinates are normalized
maxAnisotropy	unsigned int	The maximum anisotropy to use when reading from a texture
mipmapFilterMode	cudaTextureFilterMode	The interpolation method to be used when selecting levels
mipmapFilterBias	float	Value to be added when selecting mipmap levels
minMipmapLevelClamp	float	Minimum mipmap level that can be used for sampling
maxMipmapLevelClamp	float	Maximum mipmap level that can be used for sampling
disableTrilinearOptimization	int	Disables optimization for trilinear filtering
seamlessCubemap	int	Enables (1) or disables (0) filtering for cube maps

These fields closely resemble the OpenGL texture display parameters discussed in Chapter 10. The addressMode field tells CUDA how texels should be set when the application attempts to access coordinates beyond the texture's boundary. This array contains three elements that correspond to the (x, y, z) directions, and each element can be set to one of four values:

- cudaAddressModeWrap — Coordinates wrap around, so locations beyond 1.0 wrap around to 0.0
- cudaAddressModeClamp — Out-of-range values are clamped between 0.0 and 1.0
- cudaAddressModeMirror — Texels beyond the texture are set to the reflections of the corresponding in-range texels
- cudaAddressModeBorder — Texels beyond the boundary are set to the border color (defined with the borderColor field)

The filterMode field tells CUDA how to compute texels during minification and magnification. This can be set to one of two values:

- cudaFilterModePoint — computes texels using nearest-neighbor interpolation (returns the texel of the nearest point)
- cudaFilterModeLinear — computes texels using bilinear interpolation (returns the weighted average of nearby texels)

The next field, `readMode`, specifies which data type should be used for texels read from the texture. This can be set to one of two values:

- `cudaReadModeElementType` — texel values should be provided in their native data type
- `cudaReadModeNormalizedFloat` — texels should be provided as floating-point values between 0.0 and 1.0 for unsigned integers, or between –1.0 and 1.0 for signed integers

To read data from a texture, a CUDA kernel needs to provide coordinates. If the `normalizedCoords` field is set to 0, these coordinates will be unnormalized. If this field is set to 1, each coordinate's value will be expected to fall between 0 and 1.

By default, a CUDA kernel reads the same number of samples in the x and y directions. That is, sampling data is always provided in an N-by-N square. This is called *isotropic filtering* because sampling is performed the same way in every direction.

This is fine when the viewing angle is perpendicular to the texture, but if a viewer looks at an object from an angle, the object may appear to stretch in one direction and shrink in the other. To maintain resolution, the renderer should read more samples in the stretched direction. Directional sampling is called *anisotropic filtering*.

In code, anisotropic filtering is enabled by setting the `maxAnisotropy` field to a value greater than 1. This identifies the maximum ratio of the number of samples to be read in one direction divided by the number of samples to be read in the other direction. For example, if `maxAnisotropy` is set to 4, the sampler will read a maximum of four times the number of samples in one direction over the number of samples in the other direction.

If the texture has mipmap levels, each level's image has a lower resolution than the one preceding it. That is, the Level 0 image has the highest resolution, the Level 1 image has the next highest, and so on. The `mipmapFilterMode` field determines how CUDA chooses levels, and this can be set to one of two values:

- `cudaFilterModePoint` — Chooses the nearest mipmap level
- `cudaFilterModeLinear` — Uses linear interpolation between levels

When selecting a mipmap level, the value of the `mipmapFilterBias` is added to the result. A positive bias results in lower-resolution mipmap levels and a negative bias results in higher-resolution mipmap levels. By default, the bias is set to 0.0.

After mipmap filtering and bias addition, the resulting value is clamped between `minMipmapLevelClamp` and `maxMipmapLevelClamp`. By default, `minMipmapLevelClamp` is set to 0.0 and `maxMipmapLevelClamp` is set to 1000.0.

The Resource View Descriptor

The last argument of `cudaCreateTextureObject` is a `cudaResourceViewDesc`, which identifies how the data in a texture or surface should be viewed. Table 12.3 lists the fields of this structure.

Table 12.3

Fields of the cudaResourceViewDesc Structure

Field	Type	Description
`format`	`cudaResourceViewFormat`	Identifies the structure of the resource's data
`width`	`size_t`	The width of the resource view
`height`	`size_t`	The height of the resource view
`depth`	`size_t`	The depth of the resource view
`firstMipmapLevel`	`unsigned int`	The lowest mipmap level that can be accessed
`lastMipmapLevel`	`unsigned int`	The highest mipmap level that can be accessed
`firstLayer`	`unsigned int`	The lowest layer of a layered texture that can be accessed
`lastLayer`	`unsigned int`	The highest layer of a layered texture that can be accessed

The `format` field identifies the structure of the resource's data, and can take a wide variety of values. Five values for this field are given as follows:

- `cudaResViewFormatNone` — uses the resource's native format
- `cudaResViewFormatUnsignedChar4` — Four 8-bit unsigned integers
- `cudaResViewFormatUnsignedInt4` — Four 32-bit unsigned integers
- `cudaResViewFormatHalf4` — Four 16-bit floating-point values
- `cudaResViewFormatFloat4` — Four 32-bit floating-point values

In addition to mipmaps, textures can have different *layers* stacked on top of one another. This makes it possible to show a different texture when an upper layer is transparent. In the `cudaResourceViewDesc`, the `firstLayer` field sets the lowest layer that can be displayed and `lastLayer` sets the highest layer that can be displayed.

12.2.3 Reading Texture Data

After an application creates a `cudaTextureObject_t` by calling `cudaCreateTextureObject`, it can pass the texture to a kernel as a regular parameter. The kernel will receive the texture as an object of type `cudaTextureObject_t`, and it can read the texture's data using a series of functions whose names start with `tex`. Table 12.4 lists seven of these functions.

Table 12.4

Texture Access Functions (Abridged)

Function	Description
`tex1Dfetch(tex, int x)`	Reads from a 1-D texture using integer coordinate without filtering
`tex1D(tex, float x)`	Reads from a 1-D texture
`tex1DLod(tex, float x, float level)`	Reads from a 1-D texture at the given mipmap level
`tex2D(tex, float x, float y)`	Reads from a 2-D texture
`tex2DLod(tex, float x, float y, float level)`	Reads from a 2-D texture at the given mipmap level
`tex3D(tex, float x, float y, float z)`	Reads from a 3-D texture
`tex3DLod(tex, float x, float y, float z, float level)`	Reads from a 3-D texture at the given mipmap level

In addition to the listed functions, CUDA kernels can also call functions that access layered textures and cubemaps. There are also functions, such as `tex2DGrad`, that compute mipmap levels using gradient values.

Each of these functions accepts a template type that identifies the data type to be returned by the read operation. For example, the following code reads a vector containing four floating-point values (`float4`) from a `cudaTextureObject_t` named `tex`.

```
float4 texel = tex2D<float4>(tex, x, y);
```

There are no functions that update the content of textures. If you want to write to an image in a CUDA kernel, you need to create a surface. The following section explains how surfaces are created and accessed in code.

12.2.4 Surfaces

In the ch12_gaussian_blur project, the CUDA kernel reads data from a texture and filters its content. But it can't update the texture or write to another texture. To write to an image in a CUDA kernel, the image must be stored as a *surface*, which is represented by a `cudaSurfaceObject_t` structure.

Like textures, surface objects may be accessed in one, two, or three dimensions. Like textures, they can be created from CUDA arrays, CUDA mipmapped arrays, or pointers to device memory. Unlike textures, surface objects are always accessed with integer coordinates, which means that filtering isn't available. Also, features like anisotropy, mipmap filtering, and trilinear optimization aren't available when dealing with surface objects.

An application can create a surface object by calling `cudaCreateSurfaceObject`. The signature of this function is given as follows:

```
cudaCreateSurfaceObject(cudaSurfaceObject_t* surface,
  const cudaResourceDesc* resDesc)
```

The second argument points to a `cudaResourceDesc` structure, which was discussed earlier. Like textures, surfaces can be associated with CUDA arrays. To demonstrate, the following code creates and allocates a CUDA array, initializes a resource descriptor, and then creates a surface object.

```
// Create and allocate CUDA array
cudaArray_t outputImage;
cudaMallocArray(&outputImage, &channelDesc, width, height);

// Declare CUDA resource descriptor
struct cudaResourceDesc resDesc;
resDescSurf.resType = cudaResourceTypeArray;
resDescSurf.res.array = outputImage;

// Create surface object
cudaSurfaceObject_t surface;
cudaCreateSurfaceObject(&surfObj, &resDescSurf);
```

Once a surface object is created, it can be passed to a CUDA kernel as an argument. The kernel can access the surface object as a `cudaSurfaceObject_t`, and it can access the surface using functions whose name starts with `surf`. Table 12.5 lists six of these functions and provides a description of each.

Table 12.5

Surface Access Functions (Abridged)

Function	Description
`surf1Dread(surf, int x, boundaryMode)`	Reads from a 1-D surface
`surf1Dwrite(data, surf, int x, boundaryMode)`	Writes to a 1-D surface
`surf2Dread(surf, int x, int y, boundaryMode)`	Reads from a 2-D surface
`surf2Dwrite(data, surf, int x, int y, boundaryMode)`	Writes to a 2-D surface
`surf3Dread(surf, int x, int y, int z, boundaryMode)`	Reads from a 3-D surface
`surf3Dwrite(data, surf, int x, int y, int z, boundaryMode)`	Writes to a 3-D surface

In each function, the last argument is `boundaryMode`, which specifies how the kernel responds when the given coordinates lie outside the surface's boundary. This can be set to one of three values:

- `cudaBoundaryModeTrap` — out-of-range accesses cause kernel execution to fail
- `cudaBoundaryModeClamp` — out-of-range coordinates are clamped to the valid range
- `cudaBoundaryModeZero` — out-of-range reads return zero and out-of-range writes are ignored

As with the texture functions, the read functions accept a template type that identifies the type to be read from the surface. For example, the following code reads an 8-bit unsigned integer from the surface at the location given by (x, y).

```
unsigned char value = surf2Dread<unsigned char>(surface, x, y);
```

In addition to the functions listed in the table, CUDA provides functions that access surface objects with layers and cubemaps. These functions lie beyond the scope of this discussion.

12.3 Implementing the Gaussian Blur

The code in the ch12_gaussian_blur project is more complicated than any other example code in the book. It serves three main tasks:

1. Creates an OpenGL texture and a CUDA array, and then copies image data to the CUDA array.
2. Creates a texture object and surface object, and associates the surface object with the OpenGL texture.
3. Filters the content of the texture object using the Gaussian blur, and stores the result to the surface object.

This section walks through the code that performs each of these tasks. This discussion uses the term *texture* for an OpenGL texture and the term *texture object* for CUDA textures. It's important to keep these separate.

12.3.1 OpenGL Textures and CUDA Arrays

In ch12_gaussian_blur/main.cu, the `main` function calls `initTextures`, which creates the OpenGL texture and initializes the input and output CUDA arrays. Listing 12.1 presents the code for the `initTextures` function.

Listing 12.1 ch12_gaussian_blur/main.cu - the initTextures function

```
void initTextures() {

  // Create OpenGL texture
  glEnable(GL_TEXTURE_2D);
  glGenTextures(1, &texture);

  // Make texture active
  glBindTexture(GL_TEXTURE_2D, texture);

  // Read pixel data and associate it with texture
  int numChannels;
  unsigned char* imgData = stbi_load("noisy.png", &imgWidth,
    &imgHeight, &numChannels, 1);

  // Allocate memory for the texture
  glTexImage2D(GL_TEXTURE_2D, 0, GL_R8, imgWidth, imgHeight, 0,
    GL_RED, GL_UNSIGNED_BYTE, NULL);
```

Listing 12.1 ch12_gaussian_blur/main.cu - the initTextures function (Continued)

```
    // Set texture parameters
    glTexParameteri(GL_TEXTURE_2D, GL_TEXTURE_MIN_FILTER,
      GL_NEAREST);
    glTexParameteri(GL_TEXTURE_2D, GL_TEXTURE_MAG_FILTER,
      GL_NEAREST);

    // Describe the image channel
    cudaChannelFormatDesc desc =
      cudaCreateChannelDesc<unsigned char>();

    // Allocate memory for the CUDA array
    cudaMallocArray(&inputImage, &desc, imgWidth, imgHeight);

    // Copy image data to CUDA array
    cudaMemcpy2DToArray(inputImage, 0, 0, imgData,
      imgWidth * sizeof(unsigned char),
      imgWidth * sizeof(unsigned char), imgHeight,
      cudaMemcpyHostToDevice);

    // Free pixel data
    stbi_image_free(imgData);
}
```

This function starts by creating the texture descriptor, binding it to a target, and making it active. Then the function calls `stbi_load` to load image data from noisy.png.

Next, `initTextures` calls `glTexImage2D` to reserve enough memory to store the texture's image data. The last argument of `glTexImage2D` is set to `NULL` because the texture should be initially empty. Its content will be provided by the application's surface object.

Next, `initTextures` creates a `cudaChannelFormatDesc`. This specifies that the image data contains a single-channel made up of unsigned characters. Then it uses the descriptor to allocate memory for CUDA arrays whose width and height match that of the input image. After memory is allocated, `initTextures` calls `cudaMemcpy2DToArray` to copy the image data into the first CUDA array.

This application has two CUDA arrays: `inputImage` and `outputImage`. The first contains the input image data and will be associated with the texture object. The second CUDA array, `outputImage`, will be associated with the application's surface object. Once these two objects are created and configured, the CUDA kernel will read the image data in the texture object, filter it using the Gaussian blur, and write the filtered data to the surface object.

12.3.2 The Texture Object and Surface Object

After calling `initTextures`, the `main` function calls `runCuda`, which creates and configures the application's texture object and surface object. Listing 12.2 presents the code for this function.

Listing 12.2 ch12_gaussian_blur/main.cu - the runCuda function

```
void runCuda() {

  // Create the resource descriptor
  cudaResourceDesc texRes = {};
  texRes.resType = cudaResourceTypeArray;
  texRes.res.array.array = inputImage;

  // Create the texture descriptor
  cudaTextureDesc texDesc = {};
  texDesc.addressMode[0] = cudaAddressModeClamp;
  texDesc.addressMode[1] = cudaAddressModeClamp;
  texDesc.filterMode = cudaFilterModeLinear;
  texDesc.readMode = cudaReadModeNormalizedFloat;
  texDesc.normalizedCoords = 1;

  // Create the texture object
  cudaTextureObject_t texObj;
  cudaCreateTextureObject(&texObj, &texRes, &texDesc, nullptr);

  // Register the OpenGL texture
  cudaGraphicsGLRegisterImage(&cudaRes, texture, GL_TEXTURE_2D,
    cudaGraphicsRegisterFlagsSurfaceLoadStore);

  // Map the CUDA resource
  cudaGraphicsMapResources(1, &cudaRes, 0);

  // Access the CUDA array for the CUDA resource
  cudaGraphicsSubResourceGetMappedArray(&outputImage,
    cudaRes, 0, 0);

  // Create the resource descriptor for the surface object
  cudaResourceDesc surfRes = {};
  surfRes.resType = cudaResourceTypeArray;
  surfRes.res.array.array = outputImage;
```

Listing 12.2 ch12_gaussian_blur/main.cu - the runCuda function (Continued)

```
    // Create the surface object
    cudaSurfaceObject_t surfObj;
    cudaCreateSurfaceObject(&surfObj, &surfRes);

    // Invoke the kernel
    dim3 blockSize(16, 16);
    dim3 gridSize((imgWidth + blockSize.x - 1) / blockSize.x,
      (imgHeight + blockSize.y - 1) / blockSize.y);
    gaussianBlur<<<gridSize, blockSize>>>(texObj,
      surfObj, imgWidth, imgHeight);

    // Wait for the kernel to finish
    cudaDeviceSynchronize();

    // Destroy the texture and surface object
    cudaDestroySurfaceObject(surfObj);
    cudaDestroyTextureObject(texObj);

    // Unmap the CUDA resource
    cudaGraphicsUnmapResources(1, &cudaRes, 0);
}
```

This function starts by creating the structures needed to construct the texture object. The `cudaResourceDesc` structure specifies that `inputImage` will serve as the source of the texture object. Then the `cudaTextureDesc` structure sets aspects of the texture object, such as its addressing mode and filtering mode. Once these structures are set, the function calls `cudaCreateTextureObject` to create the texture object.

After creating the texture object, the function registers the OpenGL texture for CUDA access. As discussed in Chapter 11, this is accomplished with `cudaGraphicsGLRegisterImage`:

```
cudaGraphicsGLRegisterImage(cudaGraphicsResource** resource,
  GLuint image, GLenum target, unsigned int flags)
```

In the example code, the texture is identified by a descriptor named `texture`. Therefore, the following code registers the OpenGL texture with the CUDA resource:

```
cudaGraphicsGLRegisterImage(&cudaRes, texture, GL_TEXTURE_2D,
  cudaGraphicsRegisterFlagsSurfaceLoadStore);
```

The last argument indicates that CUDA will bind the texture to a surface object. But before this binding can be made, the application needs to map the resource. This is accomplished with the following code:

```
cudaGraphicsMapResources(1, &cudaRes, 0);
```

In Chapter 11, the function `cudaGraphicsResourceGetMappedPointer` was used to get a pointer to the device memory containing a vertex buffer object. For images, the function to call is `cudaGraphicsSubResourceGetMappedArray`. Its signature is given as follows:

```
cudaGraphicsSubResourceGetMappedArray(cudaArray_t* array,
  cudaGraphicsResource_t resource, unsigned int arrayIndex,
  unsigned int mipLevel)
```

This function associates a CUDA array (the first argument) with a mapped resource. The third argument identifies the index of the mapped resource, and because this application only has one, it's set to 0. Once this CUDA array (`outputImage`) has been mapped to the resource (the OpenGL texture), it can be used to create the surface object.

Next, `runCuda` creates a second `cudaResourceDesc` structure, and its fields indicate that the underlying resource is the mapped CUDA array named `outputImage`. Once this structure is created, `runCuda` calls `cudaCreateSurfaceObject` to create the surface object.

Now that the texture object and surface object have been configured, `runCuda` passes both objects to the CUDA kernel named `gaussianBlur`. This will be discussed in the next section.

After launching the kernel, `runCuda` calls `cudaDeviceSynchronize` to wait for the kernel to complete its execution. Then the function destroys the surface object with `cudaDestroySurfaceObject` and the texture object with `cudaDestroyTextureObject`. Finally, it calls `cudaGraphicsUnmapResources` to unmap the CUDA resource. Now that the resource is unmapped, the OpenGL texture can be accessed in the rendering.

12.3.3 Performing the Gaussian Blur

Before `runCuda` launches the `gaussianBlur` kernel, it sets the number of threads per thread block (`blockSize`) and the number of thread blocks in the grid (`gridSize`). Most devices support at least 256 threads per thread block, so `blockSize` is set to (16, 16), which creates 256 threads in a 16-by-16 thread block.

There's no limit on the number of thread blocks that can be created, and it's better to have too many than too few. For this reason, the x-dimension of the grid size is found by adding 15 to the image width and dividing by 16 (the x-dimension of the block size). Then the y-dimension of the grid size is found by adding 15 to the image height and dividing by 16 (the y-dimension of the block size).

Listing 12.3 presents the code that defines the coefficients of the filter and the CUDA kernel that implements the Gaussian blur.

Listing 12.3 ch12_gaussian_blur/main.cu - the gaussianBlur kernel

```
__constant__ float GAUSS_KERNEL[3][3] = {
    {1.f / 16.f, 2.f / 16.f, 1.f / 16.f},
    {2.f / 16.f, 4.f / 16.f, 2.f / 16.f},
    {1.f / 16.f, 2.f / 16.f, 1.f / 16.f}
};

__global__ void gaussianBlur(cudaTextureObject_t texObj,
  cudaSurfaceObject_t surfObj, int width, int height) {

  // Get pixel location for the thread
  int x = blockIdx.x * blockDim.x + threadIdx.x;
  int y = blockIdx.y * blockDim.y + threadIdx.y;
  if (x >= width || y >= height) return;

  // Use normalized coordinates for texture sampling
  float invW = 1.0f / width;
  float invH = 1.0f / height;
  float sum = 0.0f;

  // 3x3 Gaussian convolution
  for (int ky = -1; ky <= 1; ++ky) {
    for (int kx = -1; kx <= 1; ++kx) {
      float u = (x + 0.5f + kx) * invW;
      float v = (y + 0.5f + ky) * invH;
      float pixel = tex2D<float>(texObj, u, v);
      sum += pixel * GAUSS_KERNEL[ky + 1][kx + 1];
    }
  }

  // Convert to 8-bit and write result
  unsigned char outVal =
    static_cast<unsigned char>(fminf(sum * 255.0f, 255.0f));
  surf2Dwrite(outVal, surfObj, x * sizeof(unsigned char), y);
}
```

At the start of the code listing, `GAUSS_KERNEL` is a two-dimensional array stored in constant memory. This contains the nine values that make up the 3-by-3 Gaussian filter discussed at the start of this chapter. As explained in Chapter 4, the `__constant__` qualifier indicates that the data will be stored in read-only constant memory, which has its own dedicated cache.

The `gaussianBlur` kernel accepts four arguments: the texture object containing the image data, the surface object associated with the OpenGL texture, the image's width, and the image's height. When the kernel starts, it computes the thread's global x and y location. If x goes beyond the image width or y goes beyond the image height, the thread returns from the kernel.

The main part of the kernel consists of two loops, in which `kx` runs from -1 to 0 to 1 and `ky` runs from -1 to 0 to 1. These values are used to determine the (u, v) location of the pixel that should be used for processing.

Once the (u, v) location is determined, the kernel calls `tex2D` to read the pixel value at (u, v) and multiplies it by the appropriate value in the `GAUSS_KERNEL` array. In this manner, the two loops perform a two-dimensional dot product consisting of 3-by-3 values in a filter kernel.

The resulting value needs to be an unsigned integer between 0 and 255. Therefore, the result of the two-dimensional dot product is multiplied by 255.0, clamped to 255.0, and then converted into an unsigned integer. The last line of the kernel calls `surf2Dwrite` to update the surface object with the result.

12.4 Summary

The importance of high-speed texture processing can't be overstated. To make a game world feel responsive, applications need to display and update enormous amounts of texture data. When a character runs through a city, every surface is a texture that needs to be filtered and mapped into geometry in real time.

This chapter has shown how CUDA makes it possible to update textures before they're displayed in an OpenGL application. Unfortunately, the process isn't simple. The image data needs to be incorporated into a texture object (`cudaTextureObject_t`) and passed to a CUDA kernel for processing. The kernel reads from the texture, performs processing, and writes to a surface object (`cudaSurfaceObject_t`).

When the kernel is finished, the surface object will contain the processed pixels in a CUDA array. If this array has been associated with a texture object (by calling `cudaGraphicsSubResourceGetMappedArray`), the surface's image will be displayed in an OpenGL texture during rendering.

The example project in this chapter demonstrates how the process works. The application reads the image data into a texture object and passes it to a kernel. The kernel performs the Gaussian blur on the texture pixels and writes the data to a CUDA array associated with a surface object. Because the surface object has been mapped to an OpenGL texture, the blurred pixels will be displayed during OpenGL rendering.

To understand the code, it's important to clearly understand the difference between CUDA's texture objects and surface objects. Texture objects are optimized for reading, and can't be written to. Surface objects can be read from and written to, but they don't support the advanced reading capabilities (like sampling) that textures provide. Keep in mind that CUDA texture objects and OpenGL textures are different data structures.

Appendix A

The CUDA Driver API and Runtime Compilation

When it comes to performing low-level operations, this book has focused on the functions of the CUDA Runtime API. These functions have names like `cudaMalloc` and `cudaMemcpy`, and they're easy to use and understand. One convenience is that there's no need to explicitly select and configure a device. Once an application calls `cudaMalloc`, the runtime automatically sets up a device and context for processing.

If an application needs lower-level access to hardware, it can access the functions of the CUDA Driver API. These functions access the GPU with a `CUdevice` object and the context with a `CUcontext` object. By accessing these objects directly, applications can take advantage of advanced capabilities like multiple contexts and resource sharing.

In my opinion, the most fascinating capability of the Driver API is the ability to load and compile kernels at runtime. For normal applications, device code needs to be compiled and deployed to a device from the command line. But the Driver API makes it possible to load device code into structures called *modules* and then programmatically execute a kernel contained in the module.

The first part of this appendix presents the basic objects of the Driver API, which include devices, contexts, modules, and functions. It also presents the functions of the Driver API that transfer data between the host and device. As we'll see, these closely resemble the functions of the Runtime API.

The last part of the chapter explains how to compile code at runtime using NVRTC (Nvidia Runtime Compilation). By combining the Driver API and NVRTC, applications can compile a traditional CUDA kernel into PTX (Parallel Thread Execution), create a module from the PTX, and then launch the kernel. The project at the end of this appendix demonstrates how these operations can be performed.

A.1 The CUDA Driver API

The functions in the CUDA Driver API are similar to those we've encountered in the Runtime API, but they run at a lower level and make it possible to perform operations that directly access the GPU.

A good way to understand the Driver API is to see how it contrasts with the Runtime API. Table A.1 lists five major differences between the two.

Table A.1

Differences Between the Runtime API and Driver API

CUDA Runtime API	CUDA Driver API
Function names start with cuda	Function names start with cu
Functions declared in cuda_runtime.h	Functions declared in cuda.h
Functions provided by runtime library (libcudart)	Functions provided by driver library (libcuda)
Handles low-level tasks automatically	Requires explicit management and execution of low-level tasks
Requires kernels to be compiled and linked at compile-time	Makes it possible to launch kernels and modules dynamically at runtime

Many of the functions in the Driver API perform the same operations as the functions in the Runtime API, and most applications are better off using the Runtime API. But if a program needs to access advanced features like affinity or sharing resources with Direct3D or Vulkan, the Driver API becomes necessary.

The Driver API plays a crucial role in Appendix B, which discusses Parallel Thread Execution (PTX). PTX is Nvidia's virtual instruction set architecture, which makes it possible to write optimized code for given devices. With the Driver API, PTX code can be loaded, compiled, and executed at runtime. This is called just-in-time (JIT) compilation, which makes it possible to generate custom kernels at runtime.

A.1.1 Low-Level Operations

When you use the Runtime API, you don't need to access the GPU or perform any device initialization tasks. That is, the first time an application calls a function like `cudaMalloc`, the Runtime API performs the basic operations needed to run CUDA. This entails at least three steps.

1. Initialize the CUDA driver
2. Select the first connected CUDA-compatible device as the target
3. Create a primary context for the selected device

The Driver API provides functions that perform these steps manually. In particular, there are three functions that need to be called in order:

- `cuInit` — loads the CUDA driver, must be called before any function that accesses CUDA
- `cuDeviceGet` — provides a handle to a `CUdevice`, which represents the GPU
- `cuCtxCreate` — creates a `CUcontext` for the device, which contains all the state information associated with CUDA

As shown, the two central data structures are the `CUdevice` and the `CUcontext`. The following discussion explores functions that create, destroy, and access these data structures.

Managing Devices

An earlier discussion presented the functions in the Runtime API that managed devices, including `cudaGetDevice`, `cudaSetDevice`, and `cudaInitDevice`. The Driver API provides functions with similar names and purposes. Table A.2 lists six of them and provides a description of each.

Table A.2

Device Management Functions (Abridged)

Function	Description
`cuDeviceGetCount(int* count)`	Returns the number of connected devices that are capable of running CUDA
`cuDeviceGet(CUdevice* device, int index)`	Initializes the structure associated with the device with the given index
`cuDeviceGetName(char* name, int len, CUdevice dev)`	Provides the name of the device
`cuDeviceGetUuid(CUuuid* uuid, CUdevice dev)`	Provides the universally-unique identifier (UUID) for the device
`cuDeviceTotalMem(size_t* bytes, CUdevice dev)`	Returns the total amount of memory on the device

`cuDeviceGetAttribute(int* value, CUdevice_attribute attrib, CUdevice dev)`	Returns the value corresponding to the given attribute name

These functions are easy to understand. The first, `cuDeviceGetCount`, provides the number of connected devices capable of running CUDA. Once an application knows how many are connected, it can call `cuDeviceGet` to obtain a `CUdevice` for each device. The second argument of `cuDeviceGet` is the index of the device among the connected devices.

The next three functions in the table provide information about a `cuDevice`. That is, `cuDeviceGetName` returns the device's name, `cuDeviceGetUuid` returns its universally unique ID, and `cuDeviceTotalMem` returns the total amount of memory on the device in bytes.

The last function of the table is `cuDeviceGetAttribute`, which provides information about the given device. This performs the same operation as the `cudaDeviceGetAttribute` function discussed in Chapter 3, but uses different names. Table A.3 lists many of the attribute names that can be accessed when using the CUDA Driver API.

Table A.3

Attribute Names (Abridged)

Attribute Name	Description
`CU_DEVICE_ATTRIBUTE_MAX_THREADS_PER_BLOCK`	Maximum number of threads per thread block
`CU_DEVICE_ATTRIBUTE_MAX_BLOCK_DIM_X`	Maximum number of threads in the x-dimension of the thread block
`CU_DEVICE_ATTRIBUTE_MAX_BLOCK_DIM_Y`	Maximum number of threads in the y-dimension of the thread block
`CU_DEVICE_ATTRIBUTE_MAX_BLOCK_DIM_Z`	Maximum number of threads in the z-dimension of the thread block
`CU_DEVICE_ATTRIBUTE_MAX_GRID_DIM_X`	Maximum number of thread blocks in the x-dimension of the grid
`CU_DEVICE_ATTRIBUTE_MAX_GRID_DIM_Y`	Maximum number of thread blocks in the y-dimension of the grid
`CU_DEVICE_ATTRIBUTE_MAX_GRID_DIM_Z`	Maximum number of thread blocks in the z-dimension of the grid
`CU_DEVICE_ATTRIBUTE_MAX_SHARED_MEMORY_PER_BLOCK`	Maximum amount of shared memory available per block

CU_DEVICE_ATTRIBUTE_TOTAL_CONSTANT_MEMORY	Memory available to store constant data
CU_DEVICE_ATTRIBUTE_WARP_SIZE	The number of threads in each warp
CU_DEVICE_ATTRIBUTE_MAX_REGISTERS_PER_BLOCK	Maximum number of 32-bit registers available per block
CU_DEVICE_ATTRIBUTE_CLOCK_RATE	Clock frequency in kilohertz
CU_DEVICE_ATTRIBUTE_TEXTURE_ALIGNMENT	Required byte alignment for textures
CU_DEVICE_ATTRIBUTE_MULTIPROCESSOR_COUNT	Number of symmetric multiprocessors on the device
CU_DEVICE_ATTRIBUTE_KERNEL_EXEC_TIMEOUT	Identifies if there's a time limit for kernel execution
CU_DEVICE_ATTRIBUTE_INTEGRATED	If the host and device use integrated memory
CU_DEVICE_ATTRIBUTE_CAN_MAP_HOST_MEMORY	Identifies if host memory can be mapped to device memory
CU_DEVICE_ATTRIBUTE_COMPUTE_MODE	The device's compute mode
CU_DEVICE_ATTRIBUTE_MAXIMUM_TEXTURE1D_WIDTH	Maximum width allowed for 1-D textures
CU_DEVICE_ATTRIBUTE_MAXIMUM_TEXTURE2D_WIDTH	Maximum width allowed for 2-D textures
CU_DEVICE_ATTRIBUTE_MAXIMUM_TEXTURE2D_HEIGHT	Maximum height allowed for 2-D textures
CU_DEVICE_ATTRIBUTE_MAXIMUM_TEXTURE3D_WIDTH	Maximum width allowed for 3-D textures
CU_DEVICE_ATTRIBUTE_MAXIMUM_TEXTURE3D_HEIGHT	Maximum height allowed for 3-D textures
CU_DEVICE_ATTRIBUTE_MAXIMUM_TEXTURE3D_DEPTH	Maximum depth allowed for 3-D textures
CU_DEVICE_ATTRIBUTE_CONCURRENT_KERNELS	Specifies if the device can execute kernels concurrently
CU_DEVICE_ATTRIBUTE_ECC_ENABLED	Specifies if the device contains ECC (error-correcting code) memory
CU_DEVICE_ATTRIBUTE_PCI_BUS_ID	The PCI bus ID of the device
CU_DEVICE_ATTRIBUTE_L2_CACHE_SIZE	The size of the device's L2 cache
CU_DEVICE_ATTRIBUTE_MAX_THREADS_PER_MULTIPROCESSOR	The maximum number of threads per multiprocessor

CU_DEVICE_ATTRIBUTE_COMPUTE_CAPABILITY_MAJOR	The major number of the device's compute capability
CU_DEVICE_ATTRIBUTE_COMPUTE_CAPABILITY_MINOR	The minor number of the device's compute capability

To demonstrate, the code in the appa_device_attributes project accesses the first connected device (index = 0) and accesses three attributes: the major number of the compute capability, the minor number of the compute capability, and the maximum number of threads per thread block. Listing A.1 presents the code.

Listing A.1 appa_device_attributes/main.cu

```
// Check for errors
inline void checkErr(CUresult err, const char* file, int line) {
  if (err != CUDA_SUCCESS) {
    const char* errorMsg;
    cuGetErrorString(err, &errorMsg);
    std::cerr << "CUDA error in " << file
      << " at line " << line << ": " << errorMsg << std::endl;
    std::exit(EXIT_FAILURE);
  }
}
#define checkCudaErrors(err)
  (checkErr((err), __FILE__, __LINE__))

int main() {

  // Initialize the driver
  checkCudaErrors(cuInit(0));

  // Select the first connected device
  CUdevice device;
  checkCudaErrors(cuDeviceGet(&device, 0));

  // Access major version number of compute capability
  int ccMajor;
  checkCudaErrors(cuDeviceGetAttribute(&ccMajor,
    CU_DEVICE_ATTRIBUTE_COMPUTE_CAPABILITY_MAJOR, device));

  // Access minor version number of compute capability
  int ccMinor;
  checkCudaErrors(cuDeviceGetAttribute(&ccMinor,
    CU_DEVICE_ATTRIBUTE_COMPUTE_CAPABILITY_MINOR, device));
```

Listing A.1 appa_device_attributes/main.cu (continued)

```
  // Access max number of threads per thread block
  int maxThreads;
  checkCudaErrors(cuDeviceGetAttribute(&maxThreads,
    CU_DEVICE_ATTRIBUTE_MAX_THREADS_PER_BLOCK, device));

  // Print results
  std::cout << "Compute capability: " << ccMajor << "."
    << ccMinor << std::endl;
  std::cout << "Maximum number of threads per block: "
    << maxThreads << std::endl;
  return 0;
}
```

In addition to accessing attributes, this code demonstrates functions that check for errors. Most functions in the Driver API return a value of type `CUresult`, and if this equals `CUDA_SUCCESS`, the function completed successfully. If not, the name of the error can be obtained by calling `cuGetErrorName`, whose signature is given as follows:

```
cuGetErrorName(CUresult error, const char** str)
```

The error message can be obtained by calling `cuGetErrorString`, whose signature is given as follows:

```
cuGetErrorString(CUresult error, const char** str)
```

In addition to the functions listed in Table A.2, the Driver API also provides functions that access device properties related to remote direct memory access (RDMA) and NvSciSync attributes. These topics lie beyond the scope of this book, and will not be discussed here.

Managing Contexts

A context stores all of the state information for a device executing a given process. A device may have many associated contexts, but every context is associated with a single device.

Before an application can execute CUDA operations, it needs to access a device and create a context for that device. The primary function for creating contexts is `cuCtxCreate`, and when a context is no longer necessary, it can be destroyed by calling `cuCtxDestroy`.

Just as devices are represented by `CUdevice` objects, contexts are represented by `CUcontext` objects. Table A.4 lists `cuCtxCreate`, `cuCtxDestroy`, and other functions related to managing contexts.

Table A.4

Context Management Functions (Abridged)

Function	Description
`cuCtxCreate(CUcontext* ctx, CUctxCreateParams* params, unsigned int flags, CUdevice dev)`	Create a new context
`cuCtxDestroy(CUcontext ctx)`	Destroy the context
`cuCtxGetApiVersion( CUcontext ctx, unsigned int* version)`	Returns the API version of the context
`cuCtxGetDevice(CUdevice* device)`	Returns the device associated with the context
`cuCtxGetId(CUcontext ctx, unsigned long long* ctxId)`	Returns the context's unique ID
`cuCtxGetFlags( unsigned int* flags)`	Returns the flags for the current context
`cuCtxSetFlags( unsigned int flags)`	Sets the flags of the current context
`cuCtxGetCurrent(CUcontext* pctx)`	Returns the context bound to the calling CPU thread
`cuCtxSetCurrent(CUcontext ctx)`	Binds the given context to the calling CPU thread
`cuCtxPopCurrent(CUcontext* pctx)`	Pops the current CUDA context from the current CPU thread
`cuCtxPushCurrent(CUcontext ctx)`	Pushes a context on the current CPU thread
`cuCtxGetLimit(size_t* pvalue, CUlimit limit)`	Returns resource limits of the context
`cuCtxSetLimit(CUlimit limit, size_t value)`	Sets resource limits of the context
`cuCtxSynchronize()`	Block until the current context's tasks are completed
`cuCtxSynchronize_v2( CUcontext ctx)`	Block until the given context's tasks are completed

One recent change to CUDA is the second argument of `cuCtxCreate`. This points to a `CUctxCreateParams` structure, which makes it possible to set advanced properties of a context. This structure is defined with the following code.

```
typedef struct CUctxCreateParams_st {

    // Configures affinity
    CUexecAffinityParam *execAffinityParams;
    int                  numExecAffinityParams;

    // Configures CUDA in graphics (Cig)
    CUctxCigParam       *cigParams;
} CUctxCreateParams;
```

This structure has two pointers, and only one of them can be non-NULL. The `execAffinityParams` field relates to *affinity*, which is the ability to create a context that accesses a limited number of the GPU's streaming multiprocessors, or SMs. Affinity is important for applications that require multiple contexts executing at the same time.

The `execAffinityParams` field points to a `CUexecAffinityParam` structure, which is defined with the following code:

```
typedef struct CUexecAffinityParam_st {

    // Identifies the nature of the affinity
    CUexecAffinityType type;
    union {
        CUexecAffinitySmCount smCount;
    } param;
} CUexecAffinityParam_v1;

typedef CUexecAffinityParam_v1 CUexecAffinityParam;
```

In this structure, the `type` field identifies the nature of affinity to be employed by the context, and can be set to one of two values:

- `CU_EXEC_AFFINITY_TYPE_SM_COUNT` — create a context with a limited number of streaming multiprocessors
- `CU_EXEC_AFFINITY_TYPE_MAX` — create a context with the maximum number of streaming multiprocessors

If type is set to `CU_EXEC_AFFINITY_TYPE_SM_COUNT`, the `smCount` field must be set to the desired number of streaming multiprocessors. If type is set to `CU_EXEC_AFFINITY_TYPE_MAX`, `smCount` doesn't need to be set.

In the `CUctxCreateParams` structure, the `cigParams` field can reference a `CUctxCigParam` structure. The `Cig` stands for *CUDA in Graphics*, and this structure makes it possible to share the context with graphical toolsets. It's defined with the following code:

```
typedef struct CUctxCigParam_st {

    // Defines the graphics toolset for sharing
    CUcigDataType sharedDataType;

    // Contains the shared data
    void* sharedData;

} CUctxCigParam;
```

The `sharedDataType` field has type `CUcigDataType`, which can be set to one of two values:

- `CIG_DATA_TYPE_D3D12_COMMAND_QUEUE` — makes it possible to interact with Direct3D 12 and share resources
- `CIG_DATA_TYPE_NV_BLOB` — Makes it possible to share the context with Vulkan and other NV clients

If an application needs to interact with OpenGL graphics, the context doesn't need to be configured. But OpenGL resources must be registered and mapped to CUDA resources, and Chapters 11 and 12 discussed this in detail.

If an application doesn't need affinity or interaction with graphical toolsets, the second argument can be set to `nullptr`. This is shown in the following code:

```
// Access the first connected device
CUdevice device;
cuDeviceGet(&device, 0);

// Create a context for the device
CUcontext context;
cuCtxCreate(&context, nullptr, 0, device);
```

A host thread can only be associated with one context at a time, and this is the current context returned by `cuCtxGetCurrent`. However, the Driver API maintains a stack of contexts for each host thread, and the topmost context is current. If an application calls `cuCtxPushCurrent` with a context, that context will become current. If an application calls `cuCtxPopCurrent`, the current context is popped from the stack and the previous context will become current.

A.1.2 Modules

After creating a context, an application can send commands to the device to perform computation. Using the Driver API, this requires creating a *module*, which is a compiled unit of device code.

To understand modules, it helps to be familiar with CUDA binary files (CUBIN), parallel thread execution (PTX), and fat binaries. Chapter 2 covered these topics, and here's a quick review:

1. **CUDA Binary File (CUBIN)** — architecture-specific binary code compiled by calling nvcc with the `-cubin` flag
2. **Parallel Thread Execution (PTX)** — portable assembly language (text) compiled by calling nvcc with the `-ptx` flag
3. **Fat Binary** — combines PTX and/or CUBIN files for different architectures into a file, created by calling nvcc with multiple `-gencode` flags

With the Driver API, an application can create a module from a file by calling `cuModuleLoad`. A module can be created from text or binary data with the `cuModuleLoadData` function. Table A.5 lists these and other functions that load and unload modules.

Table A.5

Module Loading and Unloading

Function	Description
`cuModuleLoad(CUmodule* mod, const char* file)`	Load a module from a file
`cuModuleLoadData(CUmodule* mod, const void* data)`	Load a module from data
`cuModuleLoadDataEx(CUmodule* mod, const void* image, unsigned int numOptions, CUjit_option* options, void** optionValues)`	Load a module from data with additional options
`cuModuleLoadFatBinary(CUmodule* mod, const void* fatBinary)`	Load a module from fat binary data
`cuModuleUnload(CUmodule mod)`	Unload a module

These functions are easy to use. `cuModuleLoad` loads a module from a file and `cuModuleLoadData` loads a module from code. PTX code consists of text (assembly language) and CUBIN code consists of binary data.

The `cuModuleLoadDataEx` function makes it possible to configure aspects of the module's behavior. This is accomplished using *options*, which serve the same purpose as compiler flags. The function's second argument identifies the number of options that will be set, the third argument defines the names of the options, and the fourth argument sets values for the named options.

Table A.6 lists many of the module loading options that can be set in `cuModuleLoadDataEx`. The first column contains the option's name and the second column identifies the data type of the option's value.

Table A.6

Module Options

Option Name	Data Type	Description
CU_JIT_MAX_REGISTERS	unsigned int	Maximum number of registers that each thread can access
CU_JIT_THREADS_PER_BLOCK	unsigned int	Minimum number of threads per thread block
CU_JIT_WALL_TIME	float	Sets the wall clock time
CU_JIT_INFO_LOG_BUFFER	char*	Pointer to buffer to store log messages
CU_JIT_INFO_LOG_BUFFER_SIZE_BYTES	unsigned int	Number of bytes in the log buffer
CU_JIT_ERROR_LOG_BUFFER	char*	Pointer to buffer to store error messages
CU_JIT_ERROR_LOG_BUFFER_SIZE_BYTES	unsigned int	Number of bytes in the error buffer
CU_JIT_OPTIMIZATION_LEVEL	unsigned int	Defines the degree of optimization to be used for the compilation
CU_JIT_TARGET	enum	The compute capability to be targeted
CU_JIT_FALLBACK_STRATEGY	unsigned int	How to respond if the matching cubin isn't found
CU_JIT_GENERATE_DEBUG_INFO	int	Whether to generate debug information during the compilation
CU_JIT_LOG_VERBOSE	int	Whether to generate verbose log data
CU_JIT_GENERATE_LINE_INFO	int	Whether to add line number information to generated output

The CU_JIT_TARGET option is important because it identifies the compute capability of the target device. The option value must be set to the CUjit_target enumerated type, whose values take the form CU_TARGET_COMPUTE_xy where x is the major number and y is the minor number.

For example, suppose the compiler should compile the code in ptxCode and target devices with a compute capability of 12.0. In this case, the option's name is CU_JIT_TARGET and its value should be set to CU_TARGET_COMPUTE_120, as shown in the following code:

```
// Initialize variables
CUmodule module;
unsigned int numOptions = 1;
CUjit_option option = CU_JIT_TARGET;
CUjit_target optionValue = CU_TARGET_COMPUTE_120;

// Load data from PTX code
cuModuleLoadDataEx(&module, ptxCode, numOptions, &option,
  &optionValue);
```

Keep in mind that the compiler-related options only apply to code written in PTX. This is because CUBIN code has already been compiled for a given architecture.

A.1.3 Functions and Kernels

Once a module has been loaded, an application can access its functions and launch kernels. Table A.7 lists twelve functions of the Driver API that make this possible.

Table A.7

Accessing Functions and Kernels

Function	Description
cuModuleGetFunctionCount(unsigned int* count, CUmodule mod)	Returns the number of functions in the module
cuModuleEnumerateFunctions(CUfunction* functions, unsigned int numFunctions, CUmodule mod)	Accesses all functions in the module
cuModuleGetFunction(CUfunction* hfunc, CUmodule mod, const char* name)	Retrieves the function with the given name

`cuFuncLoad(CUfunction function)`	Loads a function

`cuFuncIsLoaded(`
`CUfunctionLoadingState* state,`
`CUfunction function)`
	Checks if a function is loaded

`cuModuleGetGlobal(`
`CUdeviceptr* dptr,`
`size_t* bytes, CUmodule mod,`
`const char* name)`
	Retrieves the device pointer of a global variable

`cuFuncGetAttribute(int* value,`
`CUfunction_attribute attrib,`
`CUfunction hfunc)`
	Accesses an attribute of a function

`cuFuncSetAttribute(`
`CUfunction hfunc,`
`CUfunction_attribute attrib,`
`int value)`
	Sets an attribute of the given function

`cuFuncGetName(const char** name,` `CUfunction hfunc)`	Provides the function's name

`cuFuncGetParamInfo(`
`CUfunction func,`
`size_t paramIndex,`
`size_t* paramOffset,`
`size_t* paramSize)`
	Provides the offset and size of a kernel parameter

`cuLaunchKernel(`
`CUfunction f,`
`unsigned int gridDimX,`
`unsigned int gridDimY,`
`unsigned int gridDimZ,`
`unsigned int blockDimX,`
`unsigned int blockDimY,`
`unsigned int blockDimZ,`
`unsigned int sharedMemBytes,`
`CUstream hStream,`
`void** kernelParams,`
`void** extra)`
	Launch the function as a kernel

`cuLaunchKernelEx(`
`const CUlaunchConfig* config,`
`CUfunction f,`
`void** kernelParams,`
`void** extra)`
	Launch a kernel with the given parameters

The first few entries in this table provide information about functions in a module. The first, `cuModuleGetFunctionCount`, identifies the number of functions. The second, `cuModuleEnumerateFunctions`, populates an array of `CUfunction` objects, each representing a different function in the module.

Before an application can launch a kernel, it needs to access and load the `CUfunction` corresponding to the kernel. If the kernel's name is known in advance, it can call `cuModuleGetFunction`, which will access the `CUfunction` if it's present in the module and load it automatically.

If a module needs to be accessed repeatedly, an application can load it by calling `cuFuncLoad` with the `CUfunction` pointer. It can also check if the function is loaded by calling `cuFuncIsLoaded`. In practice, this is rarely necessary.

In addition to accessing modules, an application can get a pointer to allocated device memory by calling `cuModuleGetGlobal`. This accepts the name of a variable in the module and returns a `CUdeviceptr` that can be used to access the variable's data. It also provides the number of bytes occupied by the device memory.

Applications can call `cuFuncGetAttribute` to read a function's attributes and call `cuFuncSetAttribute` to update the function's attributes. Both functions accept a `CUfunction_attribute` argument that identifies the nature of the data to be accessed. Table A.8 lists seven identifiers of function attributes.

Table A.8

Function Attributes

Attribute Name	Description
CU_FUNC_ATTRIBUTE_MAX_THREADS_PER_BLOCK	Maximum number of threads per thread block
CU_FUNC_ATTRIBUTE_SHARED_SIZE_BYTES	Size of statically-allocated shared memory required
CU_FUNC_ATTRIBUTE_CONST_SIZE_BYTES	Size of constant memory required
CU_FUNC_ATTRIBUTE_LOCAL_SIZE_BYTES	Size of local memory required
CU_FUNC_ATTRIBUTE_NUM_REGS	Number of registers required by the function
CU_FUNC_ATTRIBUTE_PTX_VERSION	Version of PTX used to compile the function
CU_FUNC_ATTRIBUTE_MAX_DYNAMIC_SHARED_SIZE_BYTES	Size of dynamically-allocated shared memory required

An example will clarify how function attributes can be accessed. The following code loads a module from module.ptx and accesses the function named `foo`. Then it calls `cuFuncGetAttribute` to get the maximum number of threads that the function can access per thread block.

```
// Load the module
CUmodule module;
cuModuleLoad(&module, "module.ptx");

// Access the function
CUfunction func;
cuModuleGetFunction(&func, module, "foo");

// Get the max number of threads
int maxThreads;
cuFuncGetAttribute(&value,
  CU_FUNC_ATTRIBUTE_MAX_THREADS_PER_BLOCK, func);
```

It's important to grasp the distinction between functions and kernels. A function is any routine that can be run on the GPU. As discussed in Chapter 4, there are two types of functions:

- **global** — the declaration starts with `__global__` and can be invoked from the host or the device
- **device** — the declaration starts with `__device__` and can be invoked from other functions running on the device

The term *kernel* refers to a global function launched from the host. To invoke a kernel with the Driver API, an application needs to call `cuLaunchKernel` or `cuLaunchKernelEx`. `cuLaunchKernel` accepts the grid dimensions (number of thread blocks per grid), block dimensions (number of threads per thread block), the amount of shared memory to be allocated, the stream in which to launch the kernel, a list of kernel parameters, and additional launch options.

For example, the following code launches a one-dimensional kernel represented by the `CUfunction` named `func`. It runs in the default stream with a grid size of `blocks_per_grid` and a block size of `threads_per_block`. The kernel parameters are stored in an array of void pointers named `kernel_params`.

```
cuLaunchKernel(func,
  blocks_per_grid, 1, 1,
  threads_per_block, 1, 1,
  0, nullptr, kernel_params, nullptr);
```

The `cuLaunchKernelEx` function serves a similar purpose, but the configuration parameters (grid size, block size, and so on) are stored in a `CUlaunchConfig` object. This accepts zero or more launch attributes, each provided as a name/value pair. These launch attributes make it possible to access advanced features that lie beyond the scope of this book.

A.1.4 Data Operations

Chapter 3 explained how to allocate blocks of device memory with functions like `cudaMalloc` and then transfer data with functions like `cudaMemcpy`. The Driver API provides its own functions that perform data operations, and they make it possible to create parameters that can be passed to a `CUfunction`. Table A.9 lists eight of these functions and provides a description of each.

Table A.9

Data Operation Functions

Function	Description
`cuMemAlloc(CUdeviceptr *dptr, size_t bytesize)`	Allocates a block of device memory of the given size
`cuMemAllocHost(void **ptr, size_t bytesize)`	Allocates a block of page-locked host memory of the given size (if supported)
`cuMemFree(CUdeviceptr dptr)`	Releases allocated memory on the device
`cuMemFreeHost(void *ptr)`	Releases page-locked host memory
`cuMemGetInfo(size_t *free, size_t *total)`	Obtains information about the free and total memory available
`cuMemcpyHtoD(CUdeviceptr dstDevice, const void *srcHost, size_t byteSize)`	Copies data from the host to device
`cuMemcpyDtoH(void *dstHost, CUdeviceptr srcDevice, size_t ByteCount)`	Copies data from the device to host
`cuMemcpyDtoD(CUdeviceptr dstDevice, CUdeviceptr srcDevice, size_t ByteCount)`	Copies data from the device to device

Pointers to device memory are represented by instances of the CUdeviceptr type. This means you don't have to know the data type of the stored values in advance. For example, the following code declares a pointer named dataMem and allocates memory for four single-precision floating-point values:

```
CUdeviceptr dataMem;
cuMemAlloc(&dataMem, 4 * sizeof(float));
```

Once a block of device memory has been allocated, applications can copy data to and from the region. The Driver API provides three different functions for transferring data, and their names identify the direction of the copy. That is, cuMemcpyHtoD copies data from the host to the device, cuMemcpyDtoH copies data from the device to the host, and cuMemcpyDtoD copies data from the device to the device.

For example, the following code declares an array of four values on the host and copies it to the device memory block identified by dataMem:

```
float dataArray[4] = { 1.0f, 2.0f, 3.0f, 4.0f };
cuMemcpyHtoD(dataMem, dataArray, 4 * sizeof(float));
```

After an application has allocated device memory and copied data to the device, it can pass the CUdeviceptr to a kernel as an input parameter. After the kernel executes, the application can copy the processed data to the host by calling cuMemcpyDtoH. The example project discussed in the next section shows how this can be accomplished.

A.1.5 Example Application — Adding Arrays

The code in the appa_add_arrays project performs the same operation as the code in the ch03_add_arrays project. That is, it sends two arrays to a kernel, which adds each pair of elements and stores the results in a third array.

The difference is that the code in appa_add_arrays uses the Driver API and loads the kernel from a PTX file named add_arrays.ptx. To be specific, the project's source code performs ten operations:

1. Initializes the driver, accesses the first connected device, and creates a context.
2. Allocates memory for three blocks of device memory: arrayA, arrayB, arrayC.
3. Copies six values from the host to arrayA and arrayB.
4. Loads a module from a PTX file named add_arrays.ptx.
5. Loads a function from the module named add_arrays.
6. Executes the add_arrays kernel with six threads.

7. Calls `cuCtxSynchronize` to wait for the kernel's execution to finish.
8. Copies processed data from the device to the host.
9. Prints the results of the addition to standard output.
10. Frees the device memory pointers, unloads the module, and destroys the context.

Listing A.2 shows how these steps are implemented in code. Keep in mind that the executable must be run in a directory containing add_arrays.ptx.

Listing A.2 appa_add_arrays/main.cu

```
int main() {
  const int N = 6;

  // Initialize the driver
  checkCudaErrors(cuInit(0));

  // Access the device
  CUdevice device;
  checkCudaErrors(cuDeviceGet(&device, 0));

  // Create the context
  CUcontext context;
  checkCudaErrors(cuCtxCreate(&context, nullptr, 0, device));

  // Allocate buffers
  CUdeviceptr arrayA, arrayB, arrayC;
  checkCudaErrors(cuMemAlloc(&arrayA, N * sizeof(float)));
  checkCudaErrors(cuMemAlloc(&arrayB, N * sizeof(float)));
  checkCudaErrors(cuMemAlloc(&arrayC, N * sizeof(float)));

  // Copy values of first input array
  float aData[N] = { 2.0f, 4.0f, 6.0f, 8.0f, 10.0f, 12.0f };
  checkCudaErrors(cuMemcpyHtoD(arrayA, aData,
    N * sizeof(float)));

  // Copy values of second input array
  float bData[N] = { 3.0f, 6.0f, 9.0f, 12.0f, 15.0f, 18.0f };
  checkCudaErrors(cuMemcpyHtoD(arrayB, bData,
    N * sizeof(float)));

  // Load module
  CUmodule module;
  checkCudaErrors(cuModuleLoad(&module, "add_arrays.ptx"));
```

Listing A.2 appa_add_arrays/main.cu (Continued)

```
    // Access the kernel
    CUfunction kernel;
    checkCudaErrors(cuModuleGetFunction(&kernel, module,
      "add_arrays"));

    // Create array of kernel parameters
    void* params[] = { &arrayA, &arrayB, &arrayC };

    // Execute kernel
    checkCudaErrors(cuLaunchKernel(kernel, 1, 1, 1, N, 1, 1, 0,
      nullptr, params, nullptr));

    // Wait for kernel execution to complete
    checkCudaErrors(cuCtxSynchronize());

    // Copy result to host
    float cData[N];
    checkCudaErrors(cuMemcpyDtoH(cData, arrayC,
      N * sizeof(float)));

    // Display result
    for (int i = 0; i < N; i++) {
      std::cout << aData[i] << " plus " << bData[i]
        << " equals " << cData[i] << std::endl;
    }

    // Free resources
    checkCudaErrors(cuMemFree(arrayA));
    checkCudaErrors(cuMemFree(arrayB));
    checkCudaErrors(cuMemFree(arrayC));
    checkCudaErrors(cuModuleUnload(module));
    checkCudaErrors(cuCtxDestroy(context));
    return 0;
}
```

When executing the kernel, the fifth parameter is set to `N` because there should be one thread for each element in the input arrays. The eighth parameter is set to `0` because no shared memory is needed.

After the kernel is called, the application calls `cuCtxSynchronize` to wait for the kernel to finish. If this isn't called, the application might start copying data to the host before the kernel has finished processing.

A.2 Nvidia Runtime Compilation (NVRTC)

The Driver API makes it easy to load modules from PTX files or compiled code. But it doesn't provide any way to compile kernel code. To make this possible, Nvidia provides the functions of the runtime compilation library, or NVRTC. They're declared in the nvrtc.h header file and can be accessed through nvrtc.lib or libnvrtc.so.

In general, working with runtime compilation consists of four steps:

1. Create a program object, represented by the `nvrtcProgram` type.
2. Compile the program object.
3. Load a module from the compiled code.
4. Select a function from the module and launch the kernel.

The preceding section explained Steps 3 and 4. This section describes Steps 1 and 2, and then presents an example that shows how runtime compilation works in practice.

A.2.1 Creating and Compiling Programs

The functions in nvrtc.h are easy to understand and use. `nvrtcCreateProgram` creates a program from C/C++ source code and `nvrtcCompileProgram` compiles the program. Then there are several functions that access the compiled result. Table A.10 lists these and other functions provided in the NVRTC library.

Table A.10

Runtime Compilation Functions

Function	Description
`nvrtcCreateProgram( nvrtcProgram *prog, const char *src, const char* name, int numHeaders, const char* const *headers, const char* const *includeNames)`	Creates a program from the given source file
`nvrtcCompileProgram( nvrtcProgram prog, int numOptions, const char *const *options)`	Compiles the program with the different options
`nvrtcDestroyProgram( nvrtcProgram *prog)`	Destroys the given program

`nvrtcGetProgramLogSize(`
`  nvrtcProgram prog,`
`  size_t *logSizeRet)`
	Determines the size of the log containing the compiler's output messages

`nvrtcGetProgramLog(`
`  nvrtcProgram prog,`
`  char *log)`
	Accesses the log containing the compiler's output messages after compilation

`nvrtcGetPTXSize(`
`  nvrtcProgram prog,`
`  size_t *ptxSizeRet)`
	Determines the size of the Parallel Thread Execution (PTX) code after compilation

`nvrtcGetPTX(nvrtcProgram prog,` `  char *ptx)`	Accesses the Parallel Thread Execution (PTX) code after compilation
`nvrtcGetCUBINSize(nvrtcProgram prog,` `  size_t *cubinSizeRet)`	Determines the size of the compiled CUDA binary (CUBIN) code after compilation
`nvrtcGetCUBIN(nvrtcProgram prog,` `  char *cubin)`	Accesses the CUDA binary (CUBIN) code after compilation
`nvrtcGetLTOIRSize(nvrtcProgram prog,` `  size_t *cubinSizeRet)`	Determines the size of the compiled LTO IR code after compilation
`nvrtcGetLTOIR(nvrtcProgram prog,` `  char *LTOIR)`	Accesses the intermediate representation (IR) code after compilation

`nvrtcAddNameExpression(`
`  nvrtcProgram prog,`
`  const char *const expr)`
	Identifies a function or variable name that may be accessed by the application

`nvrtcGetLoweredName(`
`  nvrtcProgram prog,`
`  const char *expr,`
`  const char **lowered_name)`
	Accesses the mangled name of the given expression

To present the NVRTC functions in this table, this discussion splits them into three topics:

- Creating programs
- Compiling programs
- Name mangling

Once you grasp this content, you'll be able to compile and launch kernels at runtime. You'll also be able to inject custom code into the compilation process.

Creating Programs

When I first worked with NVRTC, I found the arguments of `nvrtcCreateProgram` to be confusing. To clarify how they're used, I've listed them in order:

1. `prog` — the `nvrtcProgram` that will represent the program
2. `src` — text containing the kernel code to be compiled
3. `name` — a name that will be used to identify the program
4. `numHeaders` — the number of headers that will need to be included
5. `headers` — the text containing header code
6. `includeNames` — the names of the include files to be associated with header code

The second argument contains text embodying the code to be compiled. The third argument should be set to the name of the file that will be associated with the program. In essence, this file name will serve as an ID for the program, but `nvrtcCreateProgram` will not read the text in this file. The source code must be provided as a string in the second argument.

The fourth through sixth arguments relate to *headers*, and it may seem odd to have headers in kernel code. However, this is a fascinating capability. Headers make it possible to inject custom code (functions or data) into a kernel at runtime.

To understand how headers make it possible to inject code, it helps to look at an example. Here is a portion of the code from the appa_runtime_compile project:

```
const char* kernelSrc = R"(
#include "header.h"
extern "C" __global__
void addKernel(float *a, float *b, float *c, int n) {
  int idx = blockIdx.x * blockDim.x + threadIdx.x;
  if (idx < n)
    c[idx] = foo(a[idx], b[idx]);
)";

// Header source code
const char* headerSrc = R"(
__device__ float foo(float x, float y) {
    return x + y;
}
)";

// Header arrays
const char* headers[] = { headerSrc };
const char* includeNames[] = { "header.h" };
```

In kernelSrc, the kernel code includes header.h and calls a function named foo. But there is no header.h file in the project. Instead, there's another block of code named headerSrc, which defines a function named foo. If nvrtcCreateProgram is configured correctly, it will recognize that the code in headerSrc should be inserted where header.h is included.

In this example, there's only one header, so the numHeaders argument should be set to 1. The headers argument needs to be set to an array of strings, each containing code for a particular header. In this case, it should be set to an array whose only string is headerSrc. The includeNames argument identifies the header file names that correspond to the strings in the headers argument. If these arguments are set properly, the compiler will insert the code from headerSrc whenever header.h is included.

Compiling Programs

Once an application has created a program, it's easy to compile the code by calling nvrtcCompileProgram. The second argument identifies how many options will be set and the third argument contains a string for each option. Chapter 2 presented many of the options that can be set to configure the compiler's operation.

For example, the following code compiles prog with two arguments. The first tells the compiler to target the virtual instruction set corresponding to a compute capability of 7.5 and to store the timing result in the compiler log:

```
// Set compile options
const char* options[] = { "-arch=compute_75", "--time=-" };

// Compile program
nvrtcResult compileResult =
  nvrtcCompileProgram(prog, 2, options);
```

To access the compiler's log, three steps are required:

1. Get the log's size by calling nvrtcGetProgramLogSize.
2. Allocate memory for the log's text and the escape character, \0.
3. Read the log's text into allocated memory by calling nvrtcGetProgramLog.

An empty log will contain a newline character, which means its size will be 1. The following code calls nvrtcGetProgramLogSize and then calls nvrtcGetProgramLog if the size of the log is greater than 1.

```
// Get the size of the compilation log
size_t logSize;
checkNvrtcErrors(nvrtcGetProgramLogSize(prog, &logSize));

// Print the log if not empty
if (logSize < 2) {
  std::cout << "Log empty" << std::endl;
}
else {
  std::string log(logSize, '\0');
  checkNvrtcErrors(nvrtcGetProgramLog(prog, &log[0]));
  std::cout << "Compiler log:" << log << std::endl;
}
```

If the compilation succeeds, `nvrtcCompileProgram` will return a value of `NVRTC_SUCCESS`. By default, the kernel code will be compiled into PTX. However, the compiler will produce CUDA binary (CUBIN) code if the `-cubin` option is set, and it will produce LTO IR code if the `-ltoir` option is set.

If the compilation succeeded, the compiled result will be stored in the `nvrtcProgram` instance. An application can access the result by calling the function appropriate for the result's format:

- `nvrtcGetPTX` — Accesses PTX code (the default format)
- `nvrtcGetCUBIN` — Accesses CUBIN code (configured with the `-cubin` flag)
- `nvrtcGetLTOIR` — Accesses LTO IR code (configured with the `-ltoir` flag)

Before calling these functions, an application should obtain the size of the compiled result by calling `nvrtcGet`*`XYZ`*`Size`, where *`XYZ`* can be `PTX`, `CUBIN`, or `LTOIR`. For example, the following code calls `nvrtcGetPTXSize` to determine the size of the compiled PTX. Then it creates a string to hold the code and calls `nvrtcGetPTX` to store the PTX in the string.

```
// Get the size of the compiled PTX
size_t ptxSize;
checkNvrtcErrors(nvrtcGetPTXSize(prog, &ptxSize));

// Create a string to hold the PTX
std::string ptx(ptxSize, '\0');

// Store the PTX in the string
checkNvrtcErrors(nvrtcGetPTX(prog, &ptx[0]));
```

The code needed to access CUBIN code and LTO IR code is essentially similar. The only things that change are the function names.

Name Mangling

When compiling code to PTX, the compiler assumes that the source code is written in C++, not C. One important aspect of C++ is *name mangling*, which ensures that every symbol has a unique name.

For example, suppose a C++ class has two functions named `addNums`: one for `int`s, one for `float`s. To keep them separate, a compiler will assign each one a unique name, such as `_Z13AddNumsPi` and `_Z52AddNumsXyz`. The process of devising these strange but unique names is name mangling.

When compiling C++ to PTX, name mangling presents a problem. An application can't call a function like `addNums` if the compiler changed its name to `_Z13AddNumsPi`, and the same problem affects the function's parameters.

One way to deal with this is to tell the compiler that the kernel is written in C, not C++. In code, this is accomplished by preceding the kernel function with `extern "C"`. When the compiler sees this, it knows that the function should be compiled using C conventions instead of C++ conventions, which means no mangling.

Another option is to use the last two functions in the table: `nvrtcAddNameExpression` and `nvrtcGetLoweredName`. The first function identifies a function or variable name that the application will want to access. The second function obtains the mangled name corresponding to the name used in C++.

An example will demonstrate how these functions are used. If the function `addNums` is defined in C++, the following code will determine the mangled name assigned by the compiler:

```
// Identify a name to be accessed
nvrtcAddNameExpression(prog, "addNums");

// Perform the compilation
nvrtcCompileProgram(prog, ...);

// Get the mangled name
const char *mangledName;
nvrtcGetLoweredName(prog, "addNums", &mangledName);
```

If an application accesses a library of CUDA kernels, and there's a real possibility of having functions with identical names, then using these functions is important. But for simple cases, it's easier to prevent name mangling by preceding the kernel function with `extern "C"`. This convention will be used throughout this appendix and in the example code of Appendix B.

A.2.2 Example Application

The appa_runtime_compile project demonstrates how the functions of the CUDA Driver API and NVRTC work together. That is, it uses NVRTC to compile kernel code into PTX and then uses the Driver API to create a module from the PTX and launch the kernel. Listing A.3 presents the code in the project's main.cu source file:

Listing A.3 appa_runtime_compile/main.cu

```
// Kernel code
const char* kernelSrc = R"(
#include "header.h"
extern "C" __global__
void add_arrays(float *a, float *b, float *c) {
  int i = threadIdx.x;
  c[i] = foo(a[i], b[i]);
}
)";

// Header code
const char* headerSrc = R"(
__device__ float foo(float x, float y) {
  return x + y;
}
)";

int main() {

  // Prepare header arrays
  const char* headers[] = { headerSrc };
  const char* includeNames[] = { "header.h" };

  // Create the program
  nvrtcProgram prog;
  checkNvrtcErrors(nvrtcCreateProgram(&prog, kernelSrc,
    "runtime_compile.cu", 1, headers, includeNames));

  // Compile the program
  const char* options[] = { "-arch=compute_75", "--time=-" };
  nvrtcResult compileResult =
    nvrtcCompileProgram(prog, 2, options);
```

Listing A.3 appa_runtime_compile/main.cu (Continued)

```
  // Get the size of the compilation log
  size_t logSize;
  checkNvrtcErrors(nvrtcGetProgramLogSize(prog, &logSize));

  // Print the log if not empty
  if (logSize < 2) {
    std::cout << "Log empty" << std::endl;
  }
  else {
    std::string log(logSize, '\0');
    checkNvrtcErrors(nvrtcGetProgramLog(prog, &log[0]));
    std::cout << "Compiler log:" << log << std::endl;
  }

  // Destroy the program if compilation failed
  if (compileResult != NVRTC_SUCCESS) {
    std::cerr << "Compilation failed." << std::endl;
    nvrtcDestroyProgram(&prog);
    return 1;
  }

  // Access the compiled PTX
  size_t ptxSize;
  checkNvrtcErrors(nvrtcGetPTXSize(prog, &ptxSize));
  std::string ptxCode(ptxSize, '\0');
  checkNvrtcErrors(nvrtcGetPTX(prog, &ptxCode[0]));

  // Initialize the driver
  checkCudaErrors(cuInit(0));

  // Access the device
  CUdevice device;
  checkCudaErrors(cuDeviceGet(&device, 0));

  // Create the context
  CUcontext context;
  checkCudaErrors(cuCtxCreate(&context, nullptr, 0, device));

  // Allocate buffers
  const int N = 6;
  CUdeviceptr arrayA, arrayB, arrayC;
  checkCudaErrors(cuMemAlloc(&arrayA, N * sizeof(float)));
  checkCudaErrors(cuMemAlloc(&arrayB, N * sizeof(float)));
  checkCudaErrors(cuMemAlloc(&arrayC, N * sizeof(float)));
```

Listing A.3 appa_runtime_compile/main.cu (Continued)

```
    // Copy values of first input array
    float aData[N] = { 2.0f, 4.0f, 6.0f, 8.0f, 10.0f, 12.0f };
    checkCudaErrors(cuMemcpyHtoD(arrayA, aData,
      N * sizeof(float)));

    // Copy values of second input array
    float bData[N] = { 3.0f, 6.0f, 9.0f, 12.0f, 15.0f, 18.0f };
    checkCudaErrors(cuMemcpyHtoD(arrayB, bData,
      N * sizeof(float)));

    // Load module
    CUmodule module;
    checkCudaErrors(cuModuleLoadData(&module, ptxCode.data()));

    // Access the kernel
    CUfunction kernel;
    checkCudaErrors(cuModuleGetFunction(&kernel, module,
      "add_arrays"));

    // Execute kernel
    void* params[] = { &arrayA, &arrayB, &arrayC };
    checkCudaErrors(cuLaunchKernel(kernel, 1, 1, 1, N, 1, 1, 0,
      nullptr, params, nullptr));
    checkCudaErrors(cuCtxSynchronize());

    // Copy result to host
    float cData[N];
    checkCudaErrors(cuMemcpyDtoH(cData, arrayC,
      N * sizeof(float)));

    for (int i = 0; i < N; i++) {
      std::cout << aData[i] << " plus " << bData[i] <<
        " equals " << cData[i] << std::endl;
    }

    // Free resources
    checkNvrtcErrors(nvrtcDestroyProgram(&prog));
    checkCudaErrors(cuMemFree(arrayA));
    checkCudaErrors(cuMemFree(arrayB));
    checkCudaErrors(cuMemFree(arrayC));
    checkCudaErrors(cuModuleUnload(module));
    checkCudaErrors(cuCtxDestroy(context));
    return 0;
}
```

In this code, the name of the kernel function is `add_arrays`. When the application accesses the function with `cuModuleGetFunction`, it uses `add_arrays` as the function's name. The function's name wasn't mangled because the kernel function's declaration was preceded with `extern "C"`. If this hadn't been present, the application would need to determine the mangled name before accessing the function.

The kernel code contains a statement that includes the header.h header file, but there is no header.h in the project. To understand this, you need to see how the program was created:

```
nvrtcCreateProgram(&prog, kernelSrc, "runtime_compile.cu", 1,
  headers, includeNames)
```

The fourth argument is `1`, which tells NVRTC that the program includes one header file. The names of the header files are given in the `includeNames` array and the strings containing the header code are in the headers array. As a result, NVRTC will insert the header code from the strings when the corresponding header is included.

When the program is compiled, one of the options in `nvrtcCompileProgram` is `-arch=compute_75`, which tells the compiler to target devices that support compute capability of 7.5. A more professional application might check the device's properties first to check which compute capability is supported.

A.3 Summary

Most developers don't use the CUDA Driver API because most of its operations can be performed more easily using the CUDA Runtime API. But if you need low level access to processing contexts or devices, the Driver API provides the functions that make this possible. Also, if you want to compile and deploy code programmatically, you need to have a basic grasp of the Driver API.

The first part of this appendix discussed the fundamental structures of the Driver API. A device (`CUdevice`) represents the target GPU and a context (`CUcontext`) stores the state information for a device executing a process. Applications can manage code for a given context by creating structures called modules (`CUmodule`). A module is essentially a container of GPU code, and the Driver API makes it possible to access the functions or kernels in a module programmatically.

If a module contains CUDA code, an application can compile it at runtime by calling functions of NVRTC (Nvidia Runtime Compilation). The first step is to create a program from the code by calling `nvrtcCreateProgram`. Then the program's code can be compiled with `nvrtcCompileProgram`. If the compilation was successful, an application can access the compiled PTX code with `nvrtcGetPTX`. Once the PTX code is available, the application can create a module from it and launch it using the Driver API.

The last example project in this appendix demonstrates how the CUDA Driver API and NVRTC work together. That is, the application creates a program from a string, compiles it into PTX, creates a module from the PTX, and launches the kernel. This process requires a great deal of code, but it offers tremendous flexibility in code generation and execution.

Appendix B

Parallel Thread Execution (PTX)

This topic is not for the faint of heart. So far, this book has presented CUDA using C functions with friendly names like `cudaMemcpy` and `cudaMalloc`. Applications operate on variables with straightforward datatypes like `unsigned short`, and if you need to debug an operation, you can print messages to standard output.

These luxuries aren't available when you code with Nvidia's intermediary assembler language, called Parallel Thread Execution or PTX. The code in a *.ptx file consists of instructions like `ld`, `st`, and `clz`. Instead of operating on typed variables, they manipulate data in virtual registers.

Coding in PTX is like driving a car with a manual transmission instead of an automatic transmission. It's not simple or convenient. You have to manage the low-level details on your own, which means performing memory access without the benefit of traditional pointer operators.

But what assembly coding lacks in convenience, it makes up for in performance. If an executable has been competently coded in PTX, it will run faster and occupy less space than an executable coded in C. What's more, the GPU's strengths and weaknesses become evident as you learn its basic instructions. If you restructure your algorithm accordingly, your application will be certain to reach the loftiest of programming goals: optimal performance.

This appendix begins with a basic example that shows what PTX looks like. The next part presents directives, which guide the compiler but don't perform any processing. The third and largest part presents the instructions of PTX, divided into categories according to their function. The last part explores new capabilities of PTX, including asynchronous data transfer and the WGMMA API, which performs matrix multiplication and accumulation using warp groups.

B.1 Simple Example

Before we delve into programming details, I'd like to walk through a basic example of what PTX looks like. Chapter 3 presented a simple example project named ch03_add_arrays, which added two arrays together. The project's source file, main.cu, contains a kernel whose code is given as follows:

```
__global__ void addArrays(const float* a, const float* b,
  float* c) {

  int i = threadIdx.x;
  c[i] = a[i] + b[i];
}
```

To compile this kernel into a PTX file, the `nvcc` compiler must be invoked with the `-ptx` flag. For example, the following command compiles main.cu into main.ptx:

```
nvcc -ptx main.cu -o main.ptx
```

If this completes successfully, the compiler will write the PTX code into a file named main.ptx. Listing B.1 presents the content of this file without the comments.

Listing B.1 Compiled main.ptx

```
.version 9.0
.target sm_75
.address_size 64

.visible .entry _Z9addArraysPKfS0_Pf(
  .param .u64 _Z9addArraysPKfS0_Pf_param_0,
  .param .u64 _Z9addArraysPKfS0_Pf_param_1,
  .param .u64 _Z9addArraysPKfS0_Pf_param_2
)
{
  .reg .f32   %f<4>;
  .reg .b32   %r<2>;
  .reg .b64   %rd<11>;

  ld.param.u64      %rd1, [_Z9addArraysPKfS0_Pf_param_0];
  ld.param.u64      %rd2, [_Z9addArraysPKfS0_Pf_param_1];
  ld.param.u64      %rd3, [_Z9addArraysPKfS0_Pf_param_2];
```

Listing B.1 Compiled main.ptx (Continued)

```
    cvta.to.global.u64  %rd4, %rd3;
    cvta.to.global.u64  %rd5, %rd2;
    cvta.to.global.u64  %rd6, %rd1;
    mov.u32             %r1, %tid.x;
    mul.wide.s32        %rd7, %r1, 4;
    add.s64             %rd8, %rd6, %rd7;
    ld.global.f32       %f1, [%rd8];
    add.s64             %rd9, %rd5, %rd7;
    ld.global.f32       %f2, [%rd9];
    add.f32             %f3, %f1, %f2;
    add.s64             %rd10, %rd4, %rd7;
    st.global.f32       [%rd10], %f3;
    ret;
}
```

The structure of this code closely follows that of the original kernel. The function accepts three parameters, and its declaration is followed by a series of instructions.

At a high level, the code in this listing consists of two types of statements:

1. **directives** — statements that start with a dot and a keyword (`.version`, `.target`)
2. **instructions** — statements inside curly braces that perform operations on data (`add`, `mul`, `ld`, `mov`, and others)

Once the PTX file has been generated, the PTX assembler (`ptxas`) can assemble its code into a GPU binary (cubin) that contains Nvidia's undocumented Streaming Assembler (SASS) language. This can be accomplished with the following command:

```
ptxas main.ptx -o main.cubin
```

This appendix starts by discussing many of the directives that can be accessed in PTX. Then it explores the instructions that can be executed inside a kernel or function.

B.2 Directives

Directives guide the PTX assembler and compiler but do not directly translate into executable instructions. That is, they don't move data or operate on data. They help set up the environment and describe the program's layout.

Table B.1
PTX Directives (Abridged)

Directive	Description
`.version`	Identifies the version of the PTX instruction set
`.target`	Sets the target architecture
`.address_size`	Number of bits in a memory address
`.entry`	The kernel's entry point
`.func`	Declares a device function
`.visible`	Indicates that a kernel can be accessed externally
`.param`	Identifies a parameter to be passed to a function or kernel
`.reg`	Identifies data stored in register memory
`.local`	Identifies data stored in local memory
`.shared`	Identifies data stored in shared memory
`.const`	Identifies data stored in constant memory
`.global`	Identifies data stored in global memory
`.align`	Sets memory alignment of a variable
`.size`	Specifies the size of arrays or memory blocks

The first three directives appear at the start of every PTX file. The `.version` directive tells the assembler what version of PTX is being used and the `.target` directive identifies which GPU architecture the code is meant for. The `.address_size` directive specifies the number of bytes in each memory address, and is usually set to `64`.

B.2.1 Function Declarations

Like CUDA, PTX supports two types of function declarations. Kernels are declared with the `.entry` directive and device functions are declared with the `.func` directive. If a kernel is intended to be accessed externally, the `.entry` directive should be preceded by `.visible`.

Like a C/C++ function, a PTX function has parameters in parentheses and operations in curly braces. Kernel functions always return void, so the general PTX structure of a kernel named `foo` is given as follows:

```
.visible .entry foo(...) { ... }
```

Inside the parentheses, each parameter declaration consists of the `.param` directive followed by the parameter's data type and name. Table B.2 lists eighteen of the scalar data types supported by PTX.

Table B.2

Scalar Data Types

Data Type	Description
`.s8`	8-bit signed integer
`.s16`	16-bit signed integer
`.s32`	32-bit signed integer
`.s64`	64-bit signed integer
`.u8`	8-bit unsigned integer
`.u16`	16-bit unsigned integer
`.u32`	32-bit unsigned integer
`.u64`	64-bit unsigned integer
`.f16`	16-bit floating-point value (half)
`.f16x2`	Two 16-bit floating-point values (half)
`.f32`	32-bit floating-point value (float)
`.f64`	64-bit floating-point value (double)
`.b8`	8-bit untyped data
`.b16`	16-bit untyped data
`.b32`	32-bit untyped data
`.b64`	64-bit untyped data
`.b128`	128-bit untyped data
`.pred`	Used for conditional execution (boolean)

Most of these types are easy to understand, but the `.bN` types (`.b8`, `.b16`, `.b32`, `.b64`, and `.b128`) deserve explanation. Variables of these types contain *untyped data*, which means they can be used in bitwise operations and memory operations, but not math operations. Put simply, these variables do not contain numeric data.

PTX doesn't have a specific type for pointers. If the address size is N, pointers are given as unsigned integers with N bits. For example, if the `.address_size` directive is set to `64`, each pointer will have type `u64`.

An example will make this clear. If the kernel `foo` accepts three pointer parameters named `aptr`, `bptr`, and `cptr`, its general code could be given as follows:

```
.visible .entry foo(
  .param u64 aptr,
  .param u64 bptr,
  .param u64 cptr)
{
  ...
}
```

As discussed in Chapter 4, CUDA supports vector types that contain multiple elements of the same type, such as `char4` or `float3`. In a variable declaration, a vector type is given by a scalar type followed by `x` and the number of elements in the vector. For example, a vector containing four 16-bit unsigned integers has type `u16x4` and a vector containing three 32-bit floating-point values has type `f32x3`.

B.2.2 Variable Declarations

A function block consists of statements, and every statement ends with a semicolon. A variable declaration is a statement that consists of the following three parts:

```
<memory_type> <data_type> <name>;
```

The first part, `memory_type`, identifies the type of memory in which the variable should be stored. This can be set to one of five directives:

1. `.reg` — register memory, very fast, private to each thread
2. `.local` — local memory, used for thread-private data that exceeds register storage
3. `.shared` — shared memory, shared among threads in the same thread block
4. `.const` — portion of global memory that stores constant data
5. `.global` — global memory, slowest, accessible to all threads

As discussed in Chapter 4, local memory is a portion of global memory set aside to store thread-private data. Local memory is as slow as global memory, so it's recommended to use it only when necessary.

If a variable is declared in register memory, its name must start with the `%` character. For example, the following statement declares `sum` as a 32-bit floating-point value in register memory:

```
.reg .f32 %sum;
```

To declare multiple variables in register memory with similar names, the common name must be followed by angle brackets surrounding the number of variables to be declared. For example, the following statement declares four variables in register memory that can be accessed as %nums0, %nums1, %nums2, and %nums3.

```
.reg .f32 %nums<4>;
```

When declaring variables in other memory spaces (not in registers), the % isn't used in the name. For example, the following code declares a variable named temp in global memory.

```
.global .f32 temp;
```

Arrays can't be declared in register memory, but in other memory spaces, they're declared by following the name with square brackets and the array size. For example, the following code declares an array in shared memory named values that contains 16 unsigned integers:

```
.shared .u32 values[16];
```

By default, PTX aligns variables on a boundary equal to the variable's size. For example, .f32 variables are aligned on four-byte boundaries and .u64 variables are aligned on eight-byte boundaries.

But default alignment may cause a problem when declaring arrays. For example, consider the following statement:

```
.shared .s8 smallints[16];
```

Because the type is .s8, PTX will align this array on a single-byte boundary. This can cause an issue if a data transfer operation expects smallints to be aligned on a 4-byte or 8-byte boundary. To specify the boundary, the declaration can contain the .align directive followed by the number of bytes in the boundary.

For example, the following code specifies that smallints should be aligned on an 8-byte boundary:

```
.shared .align 8 .s8 smallints[16];
```

Alignment can usually be omitted for declarations of scalar types. The alignment should never be set to a value less than the size of the data.

B.2.3 Built-in Variables

While we're on the topic of variables, it's important to be familiar with the built-in variables that can be accessed in PTX. Chapter 4 presented the built-in variables that can be accessed in regular CUDA kernels, which store information related to thread IDs, thread block IDs, and the size of thread blocks and grids. The variables listed in Table B.3 provide the same information.

Table B.3

Built-In Variables

Built-in Variable Name	Description
`%tid.x, %tid.y, %tid.z`	Index of a thread inside a block in each dimension
`%ntid.x, %ntid.y, %ntid.z`	Number of threads inside a block in each dimension
`%ctaid.x, %ctaid.y, %ctaid.z`	Index of a block inside a grid in each dimension
`%nctaid.x, %nctaid.y, %nctaid.z`	Number of thread blocks inside a grid in each dimension

These variables are all 32-bit unsigned integers (`.u32`). As indicated by the `%` in their names, they're stored in dedicated registers.

B.3 Data Operations

After a function and its variables have been declared, many types of operations can be performed. Two particularly important operations involve transferring data from the function's input parameters into registers and then transferring processed data from registers to the function's output parameters.

These and other operations can be performed using PTX data operations. This section discusses four of them:

1. `ld` — Loads data from memory into registers
2. `st` — Stores data from registers into memory
3. `mov` — Transfer data between registers or from an intermediate value to a register
4. `cvt` — Converts data from one type to another

One frustrating aspect concerning these instructions (and all PTX instructions) is that they use different vector types than variable declarations. In an instruction, vector types start with `.v` followed by the number of elements and the underlying type. For example, `.v4.f32` is a vector containing four 32-bit floating-point values and `.v2.u64` is a vector containing two 64-bit unsigned integers.

B.3.1 Loading Data from Memory

When a thread starts executing a function, all the data is in the function's parameters, which are stored in a special region of memory called *parameter memory*. But most operations require data to be in registers, which are initially empty.

To load data into registers, functions execute the `ld` instruction. The general structure is given as follows:

```
ld.<space>.<type> <dest>, [<address>];
```

Here, `space` identifies the memory space containing the data to be transferred. If data is being loaded from parameter memory, this should be set to `param`. Otherwise, it can be set to `shared`, `const`, `local`, or `global`.

The `type` portion identifies the data type and `dest` is the register into which the data should be stored. The nature of the `address` part depends on the memory space:

1. If `space` is set to `param`, `address` should be set to the name of the function parameter.
2. If `space` is set to `shared`, `const`, `local`, or `global`, address should be set to a register containing an address in the given memory space.

An example will help make this clear. Suppose `foo` receives a parameter named `mem`. The initial declaration for `foo` could be given as follows:

```
.visible .entry foo(.param u64 mem)
```

Before the data referenced by `mem` can be accessed, the parameter' s data needs to be loaded into a register. This can be accomplished by using `ld.param` and setting the address to `mem`. The following instruction loads the `mem` address into the register named `%reg_addr`.

```
ld.param.u64 %reg_addr, [mem];
```

Now suppose `%reg_addr` contains an address of an array in global memory. The following instruction loads the first element of the array into the `%reg_value` register.

```
ld.global.f32    %reg_value, [%reg_addr];
```

If `reg_addr` contains an address in shared memory, the `global` specifier in the `ld` instruction should be replaced with `shared`. The square brackets indicate that the register contains an address, and they perform the same dereferencing operation as `*` in C.

B.3.2 Storing Data to Memory

Just as `ld` moves data from memory to registers, `st` moves data from registers to memory. It's common that the first instruction after a function's variable declarations is `ld` and the function's last instruction is `st`. The structure of the `st` instruction is essentially similar to that of `ld`:

```
st.<space>.<type> [<address>], <source>;
```

In this instruction, `address` is a register containing the memory where the data is to be stored and `source` identifies the register holding the data to be stored. `space` identifies the memory space where the data should be stored.

For example, the following code stores the data from the register named `reg_source` to the global address contained in the register named `reg_addr`:

```
st.global.f32    [%reg_addr], %reg_source;
```

An `st` instruction can't access the `param` space. It can store data in memory referenced by a parameter, but it can't change the parameter itself.

B.3.3 Moving Data

While `ld` and `st` transfer data between register memory and other memory spaces, `mov` transfers data between two registers. It can also be used to write a constant value (also called an *immediate value*) into a register.

The general structure of the instruction is given as follows:

```
mov.<type> <destination>, <source>;
```

In this instruction, `destination` must be set to a register. The `source` part can be set to a register or a numeric value. For example, the following instruction copies the content of `%reg_source` to `%reg_dest` as a 32-bit unsigned integer.

```
mov.u32 %reg_dest, %reg_source;
```

Similarly, the following instruction copies 25.0 to `%reg_dest` as a 32-bit floating-point value.

```
mov.f32 %reg_dest, 25.0;
```

The `mov` instruction becomes particularly important when dealing with built-in variables like `%tid.x`. These built-in variables are stored in special read-only registers, and need to be moved to other registers before they can be used in operations.

B.3.4 Converting Data

The `cvt` instruction is similar to `mov`, but in addition to transferring data, it converts the data from one type to another. In other words, it performs the equivalent of a cast in C and C++. The instruction's structure is given as follows:

```
cvt.<dest_type>.<src_type> <destination>, <source>;
```

This usage is straightforward to understand. It copies the data from `source` into `destination` and converts the type of the data from `src_type` to `dest_type`.

For example, the following instruction converts a 32-bit unsigned integer in `%reg_source` to a 32-bit floating-point value in `%reg_dest`.

```
cvt.f32.u32 %reg_dest, %reg_source;
```

Like the `mov` instruction, `cvt` accepts immediate values (constants) as arguments. The following instruction converts the floating-point value 100.0 to an unsigned integer and stores the converted result in `%reg_dest`.

```
cvt.u32.f32 %reg_dest, 100;
```

The `cvt` instruction becomes important when performing math operations. The instructions that execute these operations will be discussed shortly.

B.3.5 Example Application

When learning any form of assembly language, it's a good idea to start with a simple example. The code in the appb_simple_ptx project demonstrates how an application can use PTX to perform basic data operations. Using the CUDA Driver API (discussed in Appendix A), it performs nine operations:

1. Initializes the driver, accesses the device, and creates a context.
2. Allocates memory for a `CUdeviceptr` that contains a single floating-point value.
3. Creates a module from the code in simple.ptx.
4. Accesses the `simple` kernel in the module.
5. Executes the `simple` kernel and passes it the `CUdeviceptr` as a parameter.
6. Waits for the kernel to complete its execution.
7. Copies the value of the `CUdeviceptr` to host memory.
8. Prints the copied value to standard output.
9. Frees the device pointer, module, and context.

When the simple kernel executes, the PTX code performs five operations:

1. Uses `.param` to load the kernel's parameter, which points to a value in global memory.
2. Creates three registers: one to hold the generic memory address, one to hold the address in global memory, and one to hold the immediate value.
3. Uses `.ld` to load the parameter's generic address into a register.
4. Uses `.mov` to copy a floating-point value (37.0) into a register.
5. Uses `.st` to store the floating-point value to the address in global memory.

Listing B.2 shows how these steps are performed in simple.ptx. As shown, comments follow regular C++ conventions. That is, single-line comments start with `//` and multi-line comments are surrounded with `/*` and `*/`.

The first two lines set the PTX version to 7.0 and the target hardware to the Streaming Multiprocessor (SM) capability 7.0. PTX is forward compatible, so this kernel can be run on any Nvidia device since the Volta architecture, which was first provided to consumers in 2017.

The kernel declaration starts with `.visible` to indicate that it can be accessed externally, and `.entry` to indicate that it's a kernel function. It receives one parameter, and the type is set to `u64` because pointers are given as 64-bit unsigned integers.

Listing B.2 appb_simple_ptx/simple.ptx

```
// Set the version and target
.version 7.0
.target sm_70
.address_size 64

// Kernel declaration
.visible .entry simple (
.param .u64 arg
)
{
  // Register declarations
  .reg .b64 %reg_addr;
  .reg .f32 %reg_value;

  // Load address into register
  ld.param.u64 %reg_addr, [arg];

  // Move constant value to a register
  mov.f32 %reg_value, 37.0;

  // Store value to global memory
  st.global.f32 [%reg_addr], %reg_value;

  // Return from the kernel
  ret;
}
```

This code ends with `ret`, which is the equivalent of the return statement in C/C++. That is, it tells the assembler that the kernel's code has ended and that control should return to the application that called it. Every kernel written in PTX needs to end with a `ret` statement.

B.4 Math Operations

Number crunching is the reason why CUDA has attracted so much attention in the software community. For this reason, it's important to be familiar with the instructions that perform math operations. PTX provides a wealth of instructions that perform arithmetic operations, logical operations, and trigonometry operations.

Table B.4 lists instructions for basic math operations that can run on nearly all of Nvidia's hardware. The middle column presents a statement that shows how the instruction can be used. In each example, `%res` is the register that stores the operation's result.

This is an abridged listing of the math instructions available in PTX. A later discussion will present advanced instructions that perform operations like asynchronous data transfer and matrix multiply and accumulate (MMA).

Table B.4

Basic Math Instructions

Instruction	Example	Description
`add`	`add.f32 %res, %r1, %r2`	Add %r1 and %r2
`sub`	`sub.f32 %res, %r1, %r2`	Subtract %r2 from %r1
`mul`	`mul.f32 %res, %r1, %r2`	Multiply %r1 and %r2
`mul24`	`mul24.lo.s32 %res, %r1, %r2`	Returns the upper or lower 24 bits after multiplication
`div`	`div.f32 %res, %r1, %r2`	Compute %r1 divided by %r2
`rem`	`rem.u32 %res, %r1, %r2`	The remainder of %r1/%r2
`rcp`	`rcp.f32 %res, %r1`	The reciprocal of %r1
`neg`	`neg.s32 %res, %r1`	The negation of %r1
`abs`	`abs.f32 %res, %r1`	The absolute value of %r1
`min`	`min.f32 %res, %r1, %r2`	The minimum of %r1 and %r2
`max`	`max.f32 %res, %r1, %r2`	The maximum of %r1 and %r2
`mad`	`mad.f32 %res, %r1, %r2, %r3`	Multiplies %r1 and %r2, adds %r3
`mad24`	`mad24.hi.s32 %res, %r1, %r2, %r3`	Returns the upper or lower 24 bits after multiplication and addition
`fma`	`fma.f32 %res, %r1, %r2, %r3`	Fused multiplication of %r1 and %r2, addition of %r3
`sad`	`sad.s32 %res, %r1, %r2, %r3`	Computes the absolute difference between %r1 and %r2 and adds the result to %r3
`sqrt`	`sqrt.f32 %res, %r1`	Computes the square root of %r1
`rsqrt`	`rsqrt.f32 %res, %r1`	Computes the reciprocal of the square root of %r1
`lg2`	`lg2.f32 %res, %r1`	Computes the Base-2 logarithm of %r1

ex2	ex2.f32 %res, %r1	Computes 2 to the power of %r1
sin	sin.f32 %res, %r1	Computes the sine of %r1
cos	cos.f32 %res, %r1	Computes the cosine of %r1
tanh	tanh.f32 %res, %r1	Computes the hyperbolic tangent of %r1
popc	popc.b32 %res, %r1	Counts the high bits in %r1
clz	clz.b64 %res, %r1	Counts the leading zeros in %r1
brev	brev.b32 %res, %r1	Reverses the bits in %r1
bfind	bfind.s32 %res, %r1	Finds the most-significant non-sign bit in %r1

These instructions are conceptually simple to understand, but there are a few complications that arise when using them. For one thing, many of these functions are only available for specific types. For example, `mul24` and `mad24` can be used with integers, but not floating-point values.

Another issue involves *modes*. Some instructions can be followed by suffixes that constrain how the instruction is performed. For example, `add.f32` adds two values and `add.rn.f32` adds two values and rounds the result.

B.4.1 Floating-Point Operations

Most of the instructions listed in the table can be used with floating-point values, but `rem`, `mad24`, `mul24`, `popc`, and `clz` only accept integer operands. For the most part, using arithmetic instructions is easy. As an example, the following code adds the two 32-bit unsigned integers in `%r1` and `%r2`, and stores the result in `%sum`.

```
add.f32 %sum, %r1, %r2;
```

Most floating-point operations accept an optional suffix that defines the rounding mode to be used on the result. This can be set to one of four values:

- `rn` — rounds to nearest even value
- `rz` — rounds toward zero
- `rm` — rounds toward negative infinity
- `rp` — rounds toward positive infinity

For example, the following code multiplies the values in `%r1` and `%r2`, and rounds the result to the smallest integer greater than the result (toward positive infinity).

```
mul.rp.f32 %sum, %r1, %r2;
```

The smallest positive 32-bit floating-point value is 1.175×10^{38}, but many operations produce results smaller than this. To keep these tiny values from being set to 0, PTX supports *denormalized numbers*. This prevents errors caused by clipping values to zero, but may degrade performance.

The `ftz` (flush-to-zero) suffix constrains how denormalized numbers are processed. If this suffix is set, any denormalized input values will be treated as zero, and denormalized output values will be set to zero. The following instruction shows how this is used:

```
div.ftz.f32 %sum, %r1, %r2;
```

Using `ftz` may improve performance, but you should only use it if you're confident that setting tiny values to zero won't produce errors.

B.4.2 Integer Operations

When operating on integers, the result may exceed the greatest possible integer (overflow) or fall below the smallest possible integer (underflow). To prevent this from happening, the `add`, `sub`, and `mad` instructions can be followed by a `sat` (saturation) suffix that clamps the output between the maximum integer value and minimum integer value.

For example, if an application needs to add two signed 32-bit integers and clamp the result between `MININT` and `MAXINT`, it can use `add.sat`. The following code demonstrates this.

```
add.sat.s32 %res, %r1, %r2;
```

When multiplying integers, the product can occupy twice as many bits as either operand. To specify how the result should be provided, the `mul` instruction accepts one of three suffixes:

- `hi` — the result contains the upper half of the product
- `lo` — the result contains the lower half of the product
- `wide` — the result contains the entire product, which occupies twice as many bits as either operand

For example, the following instruction multiplies the values in `%r1` and `%r2`, and stores the upper half of the product in the `%res` register:

```
mul.hi.s32 %res, %r1, %r2;
```

Unlike `mul`, the `mul24` instruction only accepts integer operands. If the instruction is followed by `hi`, the result will contain the upper 24 bits of the product. If it's followed by `lo`, the result will contain the lower 24 bits of the product. For example, the following statement multiplies `%r1` and `%r2`, and stores the lower 24 bits of the product in `%res`.

```
mul24.lo.s32 %res, %r1, %r2;
```

Many applications need to perform dot products at high speed, and this is why PTX provides `mad`, `mad24`, and `fma`, which perform multiplication and addition in a single instruction. `fma` can only be used with floating-point operands, `mad24` can only be used with integer operands, and `mad` can be used with both.

When used with integers, `mad` accepts the `hi`, `lo`, `wide`, and `sat` suffixes. `mad24` accepts the `hi`, `lo`, and `sat` suffixes.

B.4.3 Bit Operations

The last four instructions in the table perform bit operations, which provide information about the arrangement of bits in a value. The first three only operate on 32-bit or 64-bit untyped values, `b32` or `b64`. `popc` returns the number of high bits (ones) in the value, `clz` returns the number of low bits (zeros) before the first 1, and `brev` reverses the bits in the operand.

An example will clarify how this works. If `%r1` and `%r2` are registers that have been declared as 64-bit unsigned types (`b64`), the following instruction will store the reversed bits of `%r1` in `%r2`.

```
brev.b64  %r2, %r1;
```

The last instruction in the table, `bfind`, only accepts 32-bit and 64-bit integers as operands. It returns a value that identifies the most significant non-sign bit.

If the input is an unsigned integer (`u32` or `u64`), `bfind` provides the position of the most significant 1. If the input is a signed integer (`s32` or `s64`), `bfind` returns the position of the most significant 0 for negative inputs and the most significant 1 for non-negative inputs.

B.5 Branching

Control flow statements make it possible to execute statements according to conditions. C and C++ provide many different control flow statements, including `if-then` statements, `for` loops, and `while` loops.

In assembly, these statements boil down to branching statements. In PTX, branching is performed using `bra` instructions, which come in two types:

- **unconditional branch** — always change the flow of execution
- **conditional branch** — change the flow of execution if a condition is met

This section discusses both types of branch statements and explains how they can be implemented in code.

B.5.1 Unconditional Branch

C++ has a `goto` statement that unconditionally transfers execution to a line following a given label. It's usually discouraged, and is only recommended when a program needs to break out of nested loops.

PTX provides a similar unconditional branch statement. Its general structure is given as follows:

```
bra <label>;
```

This tells the program to jump from its current line and execute the line following the given label. In PTX, a label is a string followed by a colon. It always starts at the beginning of the line and occupies the entire line. The following code shows how this works:

```
  bra example_label;

  ...

example_label:
  add.u32 %res, %r1, %r2;
```

In this code, the `bra` statement tells the program to skip ahead and execute the add statement following the label named `example_label`. Keep in mind that labels are never indented with surrounding code and always begin at the start of the line.

B.5.2 Conditional Branch

A conditional branch statement only transfers execution if a given condition is reached. In PTX, conditions are represented by registers that contain predicate values, which are essentially the same thing as boolean values. It's common to give these registers names that start with p, as shown in the following statement:

```
.reg .pred %p1;
```

There's no way to transfer data between predicate registers and other types of registers, so PTX provides special instructions like setp. setp sets the value of a predicate register according to the result of a comparison between two registers. The general form of a setp statement is given as follows:

```
setp.<cmp>.<type> <pred_reg>, <reg1>, <reg2>;
```

This sets the value of %pred_reg to the result of the comparison between %reg1 and %reg2. The type of comparison is determined by the cmp part of the instruction, which can be set to any of the identifiers listed in Table B.5.

Table B.5

Comparison Operators

ID	Example	Description
eq	setp.eq.s32 %p, %r1, %r2;	Sets %p to true if %r1 equals %r2
ne	setp.ne.u32 %p, %r1, %r2;	Sets %p to true if %r1 doesn't equal %r2
lt	setp.lt.s32 %p, %r1, %r2;	Sets %p to true if %r1 is less than %r2
le	setp.le.f32 %p, %r1, %r2;	Sets %p to true if %r1 is less than or equal to %r2
gt	setp.gt.s32 %p, %r1, %r2;	Sets %p to true if %r1 is greater than %r2
ge	setp.ge.s32 %p, %r1, %r2;	Sets %p to true if %r1 is greater than or equal to %r2
ltu	setp.ltu.u32 %p, %r1, %r2;	Sets %p to true if %r1 is less than %r2 (unsigned)
leu	setp.leu.u32 %p, %r1, %r2;	Sets %p to true if %r1 is less than or equal to %r2 (unsigned)
gtu	setp.gtu.u32 %p, %r1, %r2;	Sets %p to true if %r1 is greater than %r2 (unsigned)

geu	setp.geu.u32 %p, %r1, %r2;	Sets %p to true if %r1 is greater than or equal to %r2 (unsigned)

The `setp` instruction is easy to use. For example, the following command sets the `%p` register to true if the signed integer in `%reg1` is greater than the signed integer in `%reg2`.

```
setp.gt.s32 %p, %reg1, %reg2;
```

When comparing values, the `lt`, `le`, `gt`, and `ge` operators use two's-complement representation, where the highest bit identifies the sign. In contrast, the `ltu`, `leu`, `gtu`, and `geu` operators assume the values are unsigned, and they don't interpret the highest bit as a sign bit.

Once the value of a predicate register is set, it can be used in a conditional branch instruction. The general form is given as:

```
@<pred_reg> bra <label>;
```

At the start, the `@` indicates that the statement is based on a predicate, and will only execute if the `pred_reg` register is true. In this example, the program will transfer execution to `label` if the value in `pred_reg` is true. If the value in `pred_reg` is false, the statement will be ignored.

As an example, the following code sets the content of `%pred` to true if the value in `%r1` equals the value in `%r2`. Then it branches to `new_label` if the content of `%pred` is true.

```
  // Set %pred to true if %r1 = %2
  setp.gt.s32 %pred, %r1, %r2;

  // Branch to new_label if %pred is true
  @%pred bra new_label;

  ...

new_label:
  sub.f32 %res, %r1, %r2;
```

This code serves the same purpose as an `if-then` statement in C and C++. If the label precedes the conditional branch, then the code resembles the `while-do` statement in C and C++.

PTX also supports branching statements that work like the `else` part of an `if-then-else` statement. That is, the statement is executed if the predicate register contains a *false* value instead of a true value. To make this possible, the register needs to be preceded with `@!` instead of just `@`.

For example, the following code branches to `true_label` if the content of `%pred` is true. It branches to `false_label` if the content of `%pred` is false.

```
// Execute branch if true
@%pred bra true_label;

// Execute branch if alse
@!%pred bra false_label;
```

In effect, the `!` symbol negates the content of a predicate register. This is similar to its use as a negation operator in several programming languages, including C and C++.

B.5.3 Predicated Execution

The preceding discussion explained how branching statements become conditional branching statements if preceded by `@%pred` or `@!%pred`, where `%pred` is a predicate register. In PTX terminology, operands like `@%pred` and `@!%pred` are called *predicate guards.*

Branches aren't the only type of statements that can be made conditional by preceding them with predicate guards. Arithmetic, logical, and data transfer instructions can also be made conditional by preceding them with predicate guards. In PTX, this is called *predicated execution.*

For example, if `%p` is a register containing a predicate value, the following instruction only performs addition if the value in `%p` is true:

```
// Perform addition if %p is true
@%p add.f32 %sum, %r1, %r2;
```

It's important to note that some PTX statements can't be preceded by predicate guards. These include control instructions like `exit` and instructions that update predicate values like `setp`.

Predicated execution serves an important purpose. If several instructions are preceded by predicate guards, there's no need to have a conditional branch statement. This means that threads in the same warp will always process the same statements. If a conditional branch statement causes threads in a warp to process different statements, the resulting *warp divergence* can cause a significant drop in performance.

B.5.4 Conditional Selection

The C and C++ languages support the ternary operator, which sets a variable to one of two values depending on a boolean condition. For example, the following code checks if `age` is greater than or equal to 18 and sets `status` to `adult` if the condition is true and `minor` if the condition is false.

```
std::string status = (age >= 18) ? "adult" : "minor";
```

Applications can perform the same operation in PTX using the `selp` instruction. This checks the value of a predicate register and uses the result to set a destination register to one of two values. The structure of the `selp` instruction is given as follows:

```
selp.<type> <dest>, <reg1>, <reg2>, <pred_reg>;
```

If the value in `%pred_reg` is true, `%dest` will be set to `%reg1`. Otherwise, `%dest` will be set to `%reg2`.

B.6 Example Application

The appb_reverse_array project demonstrates how math, branching, and data operations can be performed in a simple PTX example. As the project's name implies, the kernel receives an array of values and returns an array in which the order of the values is reversed.

The reversal algorithm iterates through the input array and sets values of the output array to the values in reverse. In regular CUDA, this can be accomplished with the following code:

```
__global__ void reverse_array(
  const uint32_t *in_array, uint32_t *out_array, uint32_t N) {

  for (uint32_t i = 0; i < N; i++) {
    out_array[N-1-i] = in_array[i];
  }
}
```

Listing B.3 presents the code of the reverse_array.ptx file. The host code loads this as a module and then executes the `reverse_array` kernel.

Listing B.3 appb_array_reverse/array_reverse.ptx

```
// Kernel declaration
.visible .entry array_reverse (
  .param .u64 in_ptr,
  .param .u64 out_ptr,
  .param .u32 n
)
{
  // Register declarations
  .reg .u64   %in_addr, %out_addr, %in_offset;
  .reg .u64   %out_offset, %current_addr;
  .reg .u32   %n, %i, %current_val;
  .reg .u32   %tmp, %rev_idx;
  .reg .pred  %loop_done;

  // Load parameters
  ld.param.u64  %in_addr, [in_ptr];
  ld.param.u64  %out_addr, [out_ptr];
  ld.param.u32  %n,   [n];

  // Initialize loop index
  mov.u32 %i, 0;

loop_start:

  // Check if loop should end
  setp.ge.u32  %loop_done, %i, %n;
  @%loop_done bra loop_end;

  // Read next value
  mul.wide.u32  %in_offset, %i, 4;
  add.u64       %current_addr, %in_addr, %in_offset;
  ld.global.u32 %current_val, [%current_addr];

  // Reverse index = n - 1 - i
  sub.u32 %tmp, %n, 1;
  sub.u32 %rev_idx, %tmp, %i;

  // Find address in output array
  mul.wide.u32   %out_offset, %rev_idx, 4;
  add.u64        %current_addr, %out_addr, %out_offset;
  st.global.u32  [%current_addr], %current_val;
```

Listing B.3 appb_array_reverse/array_reverse.ptx (Continued)

```
    // Increment the index
    add.u32 %i, %i, 1;
    bra loop_start;

loop_end:
    ret;
}
```

The main difficulty of this code involves implementing the `for` loop. The loop begins after the `loop_start` label and ends with the `loop_done` label. After `loop_start`, the code checks if `%i` is greater than or equal to `%n`. If this is the case, execution transfers to the `loop_done` label, which leads to the kernel returning (`ret`). This check is accomplished with the following code:

```
setp.ge.u32 %loop_done, %i, %n;
@%loop_done bra loop_end;
```

If `%i` is less than `%n`, another iteration is performed. This multiplies `%i` by 4 to get the offset of the next four-byte value in the input array. Then the offset is added to the address containing the array of input values. The `ld.global` instruction stores the address of the current value into a register named `%current_val`.

The output array stores values in reverse order, so the output index equals N – i – 1, where N is the number of elements in the array and i is the index of the input array. The code uses this output index to determine the address where the next value should be stored in the output array.

Once the output address is computed, the code uses `st.global` to write the content of the `%current_val` register into the memory referenced by the output address. Then the code increments the index and performs an unconditional branch to the `loop_start` label. This is performed with the following code:

```
add.u32 %i, %i, 1;
bra loop_start;
```

This code uses the `mul.wide` instruction twice, which multiplies two values and returns a value that contains twice as many bytes as either operand. In both cases, the first operand is a 32-bit index and the second operand is an immediate value (`4`). The product is a 64-bit offset value that can be added to an address in global memory.

B.7 Advanced Features

Up to this point, this appendix has focused on performing basic operations in PTX, and the example code can run on the vast majority of Nvidia devices. The goal of this section is to present more interesting capabilities that require modern hardware. To be specific, this section looks at two advanced features:

1. **Asynchronous data transfers** — instructions that transfer data asynchronously between registers and memory, and between global memory and shared memory
2. **The WGMMA (Warpgroup Matrix-Multiply-Accumulate) API** — instructions that perform MMA operations using warp groups instead of warps

This section looks at both capabilities and shows how they can be accessed with PTX instructions. Keep in mind that the main goal is to perform matrix operations as quickly as possible, which is necessary for machine learning workloads.

B.7.1 Asynchronous Data Operations

An earlier section explained how instructions like `ld` and `st` transfer data between registers and memory. These accept suffixes that specify the nature of the data and the memory to be accessed. That is, `ld.shared.u16` loads a 16-bit unsigned integer from shared memory while `st.global.f32` stores a 32-bit floating-point value to global memory.

Regular data transfers are *synchronous*, which means the warp halts until they finish executing. But Nvidia's Ampere architecture introduced two instructions that can transfer data asynchronously:

- `st.async` — transfers data from registers into shared memory of other thread blocks (but the same cluster)
- `cp.async` — transfers data between shared memory and global memory, can be performed in bulk

These instructions may look as simple as the `ld` and `st` instructions discussed earlier, but as we'll see, they're actually rather complex. Part of the complexity involves synchronization. Because these operations are asynchronous, the application needs to wait for their completion.

Another part of the difficulty is that you need to become familiar with two new terms: *clusters* and *CTAs*. This discussion introduces these terms and then explores the `st.async` and `cp.async` instructions.

Clusters and Cooperative Thread Arrays (CTAs)

Chapter 4 explored the processing geography of Nvidia devices, which includes grids, thread blocks, and streaming multiprocessors (SMs). SMs usually behave independently, but since the release of the Ampere architecture, a pair of adjacent SMs can be combined into a processing unit called a *cluster*.

The main advantage in using clusters is distributed shared memory. That is, threads in one thread block can access the shared memory of any other thread block in the cluster. This allows much greater cooperation between thread blocks.

In PTX, thread blocks are commonly called *cooperative thread arrays*, or just *CTAs*. Some instructions accept a `shared::cta` suffix that limits the operation to the shared memory of the local thread block. Another popular suffix is `shared::cluster`, which allows the operation to access any shared memory region in the cluster.

Memory Barriers

Asynchronous instructions return immediately, allowing the program to perform other operations. When the program needs to wait for `st.async` or `cp.async.bulk` to complete, it needs to access a memory barrier (frequently shortened to *mbarrier*). An mbarrier keeps track of two values:

- **arrival count** — the number of threads expected to reach the barrier before it completes
- **transaction count** — the number of bytes that need to be transferred

Memory barriers don't use special device resources, but are 64-bit variables stored in shared memory. The following code declares a memory barrier named `mbar`:

```
.shared .align 8 .b64 mbar;
```

The `mbarrier.init.shared` instruction specifies that the given variable is a memory barrier. It accepts a parameter that identifies the number of threads that are expected. As an example, the following code initializes `mbar` as a memory barrier and sets its expected arrival count to 5:

```
mbarrier.init.shared::cluster.b64 mbar, 5;
```

In this case, the `shared::cluster` designation states that the barrier applies to all the shared memory in the cluster. This can also be set to `shared::cta`, which specifies that the barrier only applies to the local thread block.

For the `st.async` or `cp.async.bulk` instructions, the barrier needs to know how much data should be transferred before it triggers. This is set with the `mbarrier.arrive.expect_tx` instruction. For example, the following instruction configures `mbar` to trigger when 9,128 bytes of data has been transferred:

```
mbarrier.arrive.expect_tx.shared::cluster mbar, 9128;
```

After a memory barrier has been initialized and configured, an application can execute `mbarrier.wait.shared`. This instruction forces the application to wait until the barrier's arrival count reaches the expected value or the data transfer size reaches the expected value. For example, the following instruction forces the application to wait until mbar is triggered:

```
mbarrier.wait.shared::cluster mbar;
```

In the following subsections, the `st.async` and `cp.async.bulk` instructions use memory barriers like `mbar` for synchronization. The regular `cp.async` instruction uses wait groups instead.

Asynchronous Store (st.async)

As discussed earlier, `st.shared` transfers data from a register into shared memory. The new `st.async` instruction performs the same operation, but it's different in two main ways:

1. It executes asynchronously, which means the instruction returns immediately and allows the processor to perform other operations while the transfer completes.
2. Instead of being limited to the shared memory of the local thread block, it can store data to any block of shared memory in the cluster.

The structure of the `st.async` instruction is given as follows:

```
st.async.weak.shared::cluster.mbarrier::complete_tx::bytes.<type>
  [<dest_addr>], <src_reg>, [<reg_barrier>];
```

In this instruction, `type` sets the data type of the value or values to be transferred. This can be set to `b32`, `b64`, `u32`, `u64`, `s32`, `s64`, `f32`, or `f64`.

Vector types (`v2` and `v4`) can also be used. This means the `type` parameter can be set to `v4.f32` for a vector containing four `float`s or `v2.i32` for a vector containing two integers.

The last parameter identifies a register containing the address of a memory barrier in shared memory. This barrier is responsible for synchronizing the data transfer, and it should be configured to trigger when the given number of bytes is transferred.

For example, suppose an application wants to store a value from `%reg_value` to the address in shared memory stored in the `%reg_addr` register. If `%reg_mbar` stores the address of a memory barrier, the following instruction will perform the store operation asynchronously.

```
st.async.weak.shared::cluster.mbarrier::complete_tx::bytes.b32
  [%reg_addr], %reg_value, [%reg_mbar];
```

This instruction will launch the data transfer and return immediately. To wait for the operation to complete, the application will need to execute `mbarrier.wait` with the memory barrier.

Asynchronous Copy (cp.async)

Normally, transferring data from global memory to shared memory requires two operations:

1. Load data from global memory into a register (`ld.global`).
2. Store data from the register to shared memory (`st.shared`).

Ampere devices and later GPUs can accelerate data transfer with `cp.async`, which moves data directly from global to shared memory. This frees the device's registers for number crunching, and the simplest structure of this instruction has the following form:

```
cp.async.<cache>.shared.global [<dest_addr>], [<src_addr>], <size>
```

The `cache` suffix specifies how the data transfer interacts with the device's cache. This can be set to one of two values:

- `cg` — Bypasses L1 cache, attempts to store data in the L2 cache
- `ca` — Attempts to store data in the L1 and L2 caches (cache all)

For most applications, `cg` is more appropriate because data being moved to shared memory doesn't need to be written to cache. However, if the data will be accessed repeatedly by different thread blocks, the `ca` suffix will make sure it's available in L1 cache.

Unlike the ld and st instructions, cp.async doesn't have a suffix that identifies the data type. Instead, the size parameter identifies the number of bytes to be transferred from src_addr in global memory to dest_addr in shared memory. This can be set to 4, 8, or 16. For transferring float4 or int4 vectors, this should be set to 16 (128 bits).

As an example, the following instruction transfers sixteen bytes from the global memory address in reg_global to the shared memory address in reg_shared. Because of the cg suffix, the transfer will bypass the L1 cache.

```
cp.async.cg.shared.global %reg_shared, %reg_global, 16;
```

The cp.async instruction executes asynchronously, which means it returns immediately instead of waiting for the transfer to complete. This enables the application to perform other operations while the data is in transit.

Large-scale data transfers will call cp.async in a loop. To simplify management, PTX makes it possible to combine transfers into groups with the cp.async.commit_group instruction. When this is executed, all preceding ungrouped copy operations will be combined in the same group.

After defining groups of operations, an application can wait for one or more groups to complete with cp.async.wait_group. This accepts a value that identifies the number of groups that may continue to execute. For example, if the instruction is followed by 0, the application will halt until every group of operations completes.

An example will clarify how these instructions work together. The following code launches three copy operations and combines them into a group with cp.async.commit_group. Then it executes cp.async.wait_group with a value of 0, which forces the application to halt until all of the groups' operations have completed.

```
// Launch three asynchronous copy operations
cp.async.cg.shared.global ...
cp.async.cg.shared.global ...
cp.async.cg.shared.global ...

// Combine the operations into a group
cp.async.commit_group

// Wait for all groups to complete
cp.async.wait_group 0
```

While cp.async is an improvement over ld.global and st.shared, it can be frustrating to call cp.async for each individual value. To improve on this, Nvidia provides bulk asynchronous copy operations, which will be discussed next.

Bulk Asynchronous Copy (cp.async.bulk)

One of the most impressive features of Hopper devices is the presence of Tensor Memory Accelerators (TMAs), which make it possible to transfer data in bulk. Applications can perform bulk transfers between global and shared memory by executing `cp.async.bulk`. Like `cp.async`, this launches a data transfer operation that executes asynchronously. But there are three major differences:

1. The `cp.async.bulk` instruction doesn't have a maximum amount of data that can be transferred. However, the data must be aligned on a 16-byte boundary and its size must be a multiple of 16 bytes.
2. Operations launched with `cp.async.bulk` are bidirectional. That is, they can transfer data from global memory to shared memory or from shared memory to global memory.
3. Bulk operations rely on memory barriers for synchronization instead of groups. An earlier discussion explained how these barriers can be declared, initialized, and configured.

The general structure of `cp.async.bulk` is given as follows:

```
cp.async.bulk.<dst_space>.<src_space>.mbarrier::complete_tx::bytes
    [<dest_addr>], [<src_addr>], <size> [<barrier>];
```

It's important to note that `dst_space` and `src_space` can be `shared` or `global`. This is because this instruction can move data from global memory to shared memory or from shared memory to global memory. The last parameter is a register that points to a memory barrier, which was discussed earlier.

An example will clarify how this instruction can be used. Suppose `%reg_shared` contains an address in the cluster's shared memory and `%reg_global` contains an address in global memory. If the memory barrier has been configured, the following instruction will perform a bulk transfer from shared to global memory:

```
cp.async.bulk.shared::cluster.global.mbarrier::complete_tx::bytes
    [%reg_shared], [%reg_global], %reg_size, [%reg_mbar];
```

After executing this, an application can wait for the transfer to complete using the `mbarrier.wait` instruction discussed earlier. For very large operations, it's a good idea to execute `mbarrier.wait` in a loop. This ensures that the application can perform other tasks while waiting for the asynchronous operation to finish.

B.7.2 The Warp Group Matrix Multiply-Accumulate (WGMMA) API

Chapter 5 explained how Nvidia's WMMA (Warp Matrix Multiply-Accumulate) API makes it possible to multiply and add matrices at high speed by combining all 32 threads in a warp. To improve on this, the Hopper architecture introduced the WGMMA (Warp Group Matrix Multiply-Accumulate) API. This further accelerates matrix multiply-and-add operations by combining all the threads in a *warp group* (combination of four warps) together.

As a reminder, the goal of a matrix multiply-and-accumulate operation is to multiply Matrix A and Matrix B and add the product to Matrix C. The result is stored in Matrix D. When using the WMMA API, Matrix A and Matrix B need to be loaded into registers. But WGMMA can access shared memory directly. To facilitate this, the application needs to provide a descriptor object that specifies where and how the matrix is stored.

The central instruction of the WGMMA API is `wgmma.mma_async`. Its general structure is given as follows:

```
wgmma.mma_async.sync.aligned.<shape>.<typeD>.<typeA>.<typeB>.
  mat_d, mat_a_desc, mat_b_desc, mat_c;
```

The `shape` suffix identifies the dimensions of the matrices in the operation. The size of Matrix A is given as M-by-K, the size of Matrix B is K-by-N, and the size of Matrix C is M-by-N. The `shape` suffix has the general form m<dimM>n<dimN>k<dimK>, where `dimM`, `dimN`, and `dimK` correspond to the appropriate matrix dimensions.

An example will help make this clear. Suppose Matrix A is 64-by-16, Matrix B is 16-by-256, and Matrix C is 64-by-256. To perform the MMA operation, the `shape` suffix will need to be set to `m64n256k16`. The supported shapes depend on the data types of the matrices' elements. For floating-point matrices, M must be 64, K must be 16, and N can be set to any multiple of 8 from 8 to 256.

The next three suffixes identify the types of the elements that make up Matrix D, Matrix A, and Matrix B. For floating-point matrices, these suffixes can be set to several types including `f16`, `bf16`, and `f32`. The instruction also supports integer types like `s32` and bit types like `b32`. The data type of Matrix D will also serve as the type of Matrix C, so if the suffixes are set to `f32.f16.f16`, the multiplication will involve half-precision values and further operations will involve single-precision values.

A single register can't hold the elements of Matrix D or Matrix C, so the `mat_d` and `mat_c` parameters must be set to several registers in curly braces. The elements of Matrix A and Matrix B can be stored in registers or in shared memory. If a matrix is stored in shared memory, the corresponding parameter must be set to a 64-bit value called a *descriptor*.

Each matrix descriptor must provide three pieces of information:

1. The base address of the matrix in shared memory
2. The stride of the matrix (byte offset between rows)
3. Swizzle mode (how the data is interleaved)

Like `st.async` and `cp.async`, the `wgmma.mma_async` instruction is asynchronous and returns immediately. This means the application needs a way to halt until the operation completes. The first synchronization step is to follow one or more MMA instructions with `wgmma.commit.group.sync.aligned`. This specifies that preceding ungrouped operations belong to the same group.

After creating the groups, the application can execute `wgmma.wait_group.sync.aligned N`, where `N` is the number of operations that are allowed to continue executing. If `N` is zero, the instruction will force the application to wait until every MMA operation completes.

B.8 Summary

Assembly language tends to scare developers, but if you have a solid grasp of regular CUDA, PTX won't pose too much difficulty. The instruction syntax may seem odd, but the fundamental concepts are still the same. An application starts by declaring variables allocated from the different memory spaces. Then threads execute instructions that access this data.

The first part of this appendix discusses the directives that provide information to the assembler and the data operation instructions. The `ld` instruction loads data from memory into a register, the `st` instruction stores data from a register to memory, and the `mov` instruction moves data between registers. The `cvt` instruction converts data from one type to another.

The next two sections discuss math operations and branching. The math operations are straightforward to understand, and when processing integers, it's important to be aware of underflow, overflow, and saturation. Branching selectively executes instructions according to values stored in predicate registers.

The last part of the chapter explores two advanced features that have been recently added to CUDA. The first involves asynchronous transfer operations, which make it possible to transfer data while other operations execute. The second feature relates to the WGMMA API, which makes it possible to perform high-speed matrix multiply and accumulate operations using threads belonging to a warp group.

Index

F

G

H

I

J

K

L

M

N

O

P

T

U

V

W

Z

Made in United States
Orlando, FL
04 November 2026

90207383R00199